Against Frog Dissection

Robert Rosenberger

Against Frog Dissection

The Materiality Argument

palgrave
macmillan

Robert Rosenberger
Jimmy and Rosalynn Carter School
of Public Policy
Georgia Institute of Technology
Atlanta, GA, USA

ISBN 978-3-032-14345-7 ISBN 978-3-032-14346-4 (eBook)
https://doi.org/10.1007/978-3-032-14346-4

This work was supported by Georgia Institute of Technology.

This Palgrave Macmillan imprint is published by the registered company Springer Nature Switzerland AG
The registered company address is: Gewerbestrasse 11, 6330 Cham, Switzerland

AI statement: No AI-generated text was used in the writing of this book.

Contents

CHAPTER 1

Introduction: The Materiality Argument Against Frog Dissection

Abstract As a contribution to the criticism of precollege animal dissection, I develop what can be referred to as the materiality argument. My suggestion is that by approaching both the dissection activity itself as well as its simulated alternatives as concrete material practices, we can shake off some of the assumptions about these activities that tend to lead people to favor the dissection exercise. I argue that if we can find ways to shed these biases, then we could better evaluate dissection's actual educational advantages, limitations, and problems. In addition, we could further unlock the yet unmet potential of dissection simulation. This chapter serves as an introduction to the project of developing the materiality argument against dissecting animals in the classroom.

Keywords Frog dissection · Educational technologies · Biology education · Simulation · Animal welfare · Humane education · Technological mediation · Philosophy of technology · Postphenomenology

A review of Digital Frog 2, an educational frog dissection computer program, was posted on the website for *MacWorld* magazine back in 2000. The simulation received high marks, a four out of five, represented in little computer mouse icons. The author touted its educational

R. Rosenberger, *Against Frog Dissection*,
https://doi.org/10.1007/978-3-032-14346-4_1

virtues, and described it to be "as close to the real experience of dissection as a student can get—without actually touching a frog."[1] The "AT A GLANCE" summary at the end of the review put it this way:

Pros

- Comprehensive ecological, anatomical, and biological information

Cons

- Not the real thing

I've always thought this review encapsulates much of what is at issue in debates over animal dissection in schools. It captures the way that alternatives to dissection, such as computer simulations, provide a good—or even potentially in some ways better—educational experience. (As we'll see, the empirical studies on learning outcomes largely find alternatives to be equal, and sometimes even superior, to animal dissection). The review hints that it remains morally fraught to use dead animals as pedagogical tools. And it captures as well that, despite all this, many remain hung up on the dead frog's status as a real thing and the purportedly special pedagogical value that comes with that.

This makes intuitive sense. The computer simulation and the dead frog-for-dissection *are not the same thing*. When taken up as educational tools, we shouldn't expect the two to provide the same experience. As a classroom activity, frog dissection is distinctive in many ways. It has a hands-on quality. It has a visceral impact. It's memorable. It's strange. It's smelly. It is also morally problematic. Many critics have argued over the years that it is unethical to kill what can amount to millions of frogs yearly for this activity, and millions more additional animals for educational dissection, especially when these same lessons can be taught adequately through other means.

So, we have a dilemma. Frog dissection, along with other forms of animal dissection, has of course become commonplace in some parts of the world. Even more, in these places this activity is deeply rooted within precollege biology curricula, defended by many biology teachers, and

[1] See: "The Digital Frog 2" reviewed by Adrienne Robillard, *Macworld* magazine, Jul 31st, 2000. https://www.macworld.com/article/158574/15reviewsfrog.html.

established within the larger culture as an expected and even iconic part of growing up. Nevertheless, vocal critics, including ethicists, environmentalists, animal advocates, and concerned students and parents continue to raise objections.

One central line of advancement within the debate has been the development of dissection simulations, including computer simulations and dissectable physical models. These simulations tend to take a specific form: the direct mimicry of the frog dissection activity. That is, these simulations often set out to reproduce the experience of dissection as closely as they can. The implicit moral argument is clear: if a simulation could prove to be an adequate substitute for animal dissection, then we should make this replacement since we could retain the educational value without needing to kill any animals in the process. Of course, defenders of frog dissection counter that no simulation can adequately reproduce the same experience of the "real" thing, with its hands-on character, complexity, multisensoriality, and vividness.

My suggestion is that the above captures much of the general contemporary discussion on educational dissection, especially in those regions of the world where it is common. On one side, defenders tout the *educational value* of frog dissection, often emphasizing its hands-on and visceral quality. On the other, critics make *moral arguments* against the practice, and they promote alternatives. Of course, there are many other more sophisticated points and arguments made, especially within the activist and academic literature. However, insofar as this captures much of the discussion as it occurs in classrooms, homes, general public discourse, and other spaces, I suggest that *the deck is stacked against alternatives* to animal dissection.

That is, the state of the discussion on educational dissection puts the efforts of critics at a disadvantage. I suggest that in those regions where the practice of dissecting frogs and other animals in school is commonplace, this activity has the status of a kind of cultural default. With dissection as the expectation, it is often regarded as an educational ideal, something which any proposed alternative must attempt to copy. In this scenario, simulated alternatives are set up to always fall short. If simulations must be evaluated mainly in terms of how well they replicate the exact experience of cutting up a dead frog, then the debate is already lost.

This all raises the question: What does it take to think about classroom dissection differently? There are some main things that get associated with the frog dissection activity, such as the idea that it is real, or that it's natural, or that it's a scientific experiment, and that simulations are instead virtual and artificial, and that they must mainly attempt to copy the original, and so on. How do these things all come together to discourage the critical analysis of frog dissection? And how do they get in the way of innovating dissection simulations?

We need to somehow change the discussion. It seems true that dissection simulations cannot duplicate the same exact visceral experience as dissecting a dead frog. It is possible that they will never be able to do it. However, they shouldn't have to. This whole arrangement—in which the only way we could ever consider removing the practice of dissection in precollege schooling is by replacing it with something identical—needs to be overturned. How do we displace this prevailing way of thinking?

My suggestion is that one possible way forward is through the reconsideration of all of these elements of this discussion in terms of their concrete material circumstances. Rather than as "ideals" or "assumed defaults," it may be helpful to recast all of the things at issue within this discussion in terms of their status as physical things. Of course, the computer and physical simulations are material objects, things made by people. However, the frog dissection activity itself can also be approached this way. This could enable the application of insights from the philosophy of technology and social theory.[2] It could make it possible to reconsider the preconceptions and assumptions that undergird much of the support for the use of animal dissection in precollege biology classes. Let's refer to this approach as "the materiality argument."

The materiality argument against frog dissection proceeds in two parts. The first part consists of the identification of the set of ideas that are implicitly associated with one another and which work together to support the continued use of animal dissection in schools. Following others, let's refer to these various assumptions and preconceptions as "the dissection paradigm."[3] In what follows, we'll elaborate on these initial

[2] In this way, this book can be also understood as another connection made between the fields of philosophy of technology and animal studies (e.g., Haraway 2008; Wellner 2017; Shew 2017).

[3] For example, as Jan Oakley puts it, "The current culture of North American school science continues to operate within a dissection paradigm; that is, dissections remain commonplace" (2013, 4). This quote comes from Oakley's pamphlet on these issues, entitled *Animal Dissections in Schools: Life Lessons, Alternatives, and Humane Education*, one of my favorite resources in this literature (2013). It's freely available online here: www.animalsandsociety.org/research/digital-archive/.

observations about the state of the discussion on classroom dissection. We can think further about what ideas get smuggled in, and what gets discreetly ignored, under this paradigm of thinking. What else gets tacitly assumed when the frog dissection activity is cast as something merely "real" or "natural," or as an in-class experiment, or as an educational ideal? We can also explore the wider landscape of this discussion, which includes things such as "student choice" laws that enable students to opt out of this activity, as well as the ongoing development of alternatives to dissection which largely operate within a framework of "simulating" animal dissection.

Once the general contours of the dissection paradigm have been outlined, they can be potentially reconsidered. This brings us to the second part of the materiality argument: reframing the frog dissection activity as a concrete material practice. In particular, *I suggest that we attempt to reconceive the dead frog itself as a form of technology*. If for some people this sounds counterintuitive, then that's exactly the point. The goal here is exactly to expose and critically analyze some of the key intuitions that for many people often go otherwise unexamined, intuitions that I suggest do a lot of work in guiding the discussion around educational frog dissection.

If we can find ways to think about the dead frog in its role as a technology within this learning activity, then that could help to emphasize how it is something that people are using for particular purposes in the classroom. Rather than only intuitively associating the frog with its status as a part of nature, or a specimen, or an object of investigation, we could additionally call attention to the ways it has been altered by human hands in the process of presenting it to students as a frog-for-dissection. We could think about the frog as a form of what philosophers of technology call "technological mediation." That is, we can approach the frog-for-dissection as something that comes between a student and their experience of the world and reshapes that experience in a non-neutral manner, just as the usage of any technology does. Rather than think of the frog-for-dissection as the real thing, and the alternatives as the human-made technologies, we can instead recognize both the dead frog and its alternatives as classroom tools. This sets us up to compare them against one another in a fresh way.

My suggestion is that if we can truly approach both the frog-for-dissection as well as its alternatives as technologies, then we are primed to think past the dissection paradigm. This prompts two productive directions for evaluating these educational tools. First, if we can shed the assumptions of the dissection paradigm, then we can highlight the often overlooked educational disadvantages of the frog dissection activity, as well as the educational problems that accompany it. And second, with the expectations of the dissection paradigm set aside, we can brainstorm ways to expand dissection alternatives. If we are released from the assumption that a dissection simulation must mainly attempt to replicate the experience of cutting up a dead frog, then it may be possible to take advantage of the distinctive things that may be afforded by simulations, including computer simulations. This could open up new educational possibilities.

In this way, while the philosophical reflections in what follows work in the service of ethics and animal advocacy, this is not an ethics book. We'll use ideas from the fields of philosophy of technology and Science & Technology Studies to think about the objects and practices involved in classroom dissection. This can help us uproot the unconscious assumptions attached to them. It can help identify the under-recognized educational limitations and problems of the frog dissection activity. And it can help us find new directions for simulation development. Of course, ethical questions are inextricably associated with the debate over educational dissection. So, while the materiality argument itself is not an ethical argument, it can be used to support the moral and environmental critiques of animal dissection in the classroom.

In what follows, Chapter 2 explores the details of the dissection paradigm. This chapter reviews some of the practices, policies, research, and debates regarding dissection in schools. And we'll unpack what it means to associate frog dissection with a number of key ideas, including its hands-on quality, its status as a real rather than virtual thing, and its posture as a scientific lab experiment. While all of these elements and more are quite intuitive, they also can obscure our thinking on some aspects of frog dissection and its alternatives.

Chapter 3 proposes some ways to rethink frog dissection and its alternatives that can help us to shed the preconceptions of the dissection paradigm. Ideas from the fields of philosophy of technology and Science and Technology Studies can be used to reframe our thinking on these issues. In particular, I propose that we approach both the frog dissection activity and its alternatives as forms of technological mediation, that is, as relationships with specific material tools that reshape our experience of the world. Doing so can help to reveal the contingency and specificity of the dissection paradigm, and open the door for thinking about both frog dissection and its alternatives otherwise.

And finally in Chapter 4, when moved out from under the shadow of the dissection paradigm, we can revisit frog dissection and its alternatives in a new light. With those assumptions identified and set aside, we'll do two things: (1) consider the often under-recognized educational drawbacks and problems of the frog dissection activity, and (2) brainstorm the under-explored ways that dissection simulations could be expanded when unshackled from the expectation that they should mainly aim to mimic the experience of dissecting a dead animal.

A brief concluding fifth chapter reiterates these main points of this materiality argument against frog dissection developed across the book, and considers how this perspective provides support for the moral and environmental criticisms of this practice.

One caveat to note is that the arguments of this book are intended to apply to precollege education. That is, they are directed at the practices of animal dissection as they take place in high schools and middle schools, or K-12 education, among the other names of this era of education in young people's lives. There is of course an ongoing debate as well over the use of animal dissection activities in college and professional specialist education, including medical and veterinary school. Ideas from the materiality argument could be applied to these other contexts, but that would be a different project.

Another thing to note is that in what follows I focus specifically on the practice of dissecting frogs in schools. However, I intend these arguments to apply to the dissection of animals in general. So, while in what follows we'll go into detail about the practices and problems and debates around frog dissection, these points should be understood to be applicable to animal dissections in schools more broadly.

References

Haraway, Donna J. 2008. *When Species Meet*. Minneapolis: Minnesota University Press.

Oakley, Jan. 2013. *Animal Dissection in Schools: Life Lessons, Alternatives and Humane Education* [Policy paper.]. Ann Arbor, MI: Animals and Society Institute. Available online at: www.animalsandsociety.org/research/digital-archive/

Shew, Ashley. 2017. *Animal Constructions and Technological Knowledge*. Lanham: Lexington Books.
Wellner, Galit. 2017. "Do Animals Have Technologies?" *Studia Phaenomenologica* 17: 265–282.

CHAPTER 2

The Dissection Paradigm

Abstract In those places around the world where precollege animal dissection remains common, it is possible to identify several pervasive preconceptions that tend to bias people in favor of acceptance of this practice and against simulated alternatives. We can refer to this pattern of thinking as "the dissection paradigm." After a review of the contemporary situation, this chapter outlines several of the dissection paradigm's core features. These include the idea that animal dissection is "real," and alternatives are merely "virtual," and that dissection represents an educational ideal that alternatives must mainly strive to copy. These include as well dissection's status as a "hands-on" science-like "experiment." While initially intuitive, these preconceptions can also be shown to be complex, contingent, and fraught.

Keywords Frog dissection · Educational technologies · Biology education · Simulation · Animal welfare · Humane education · Technological mediation · Philosophy of technology · Postphenomenology

It is my suggestion that in those parts of the world where frog dissection is common in childhood and young adult education, we can observe some patterns of thinking. These patterns are composed not only of explicit opinions and arguments, but also, importantly, many specific associations.

R. Rosenberger, *Against Frog Dissection*,
https://doi.org/10.1007/978-3-032-14346-4_2

That is, I suggest that much of the relevant thinking on issues of educational frog dissection is happening in terms of intuitions, in terms of unreflexive and automatic preferences, and in terms of what ideas get associated together. These kinds of patterns can be observed everywhere in discussions on frog dissection, e.g., within journalism and other coverage of these topics, teacher training, in the design of alternatives, in relevant legislation, and the official statements of teaching organizations. I will refer to these patterns of thinking here as "the dissection paradigm."

My claim is that the dissection paradigm has the effect of biasing people toward an acceptance of the frog dissection activity over alternatives. The various intuitions, and framings, and unreflexive associations of the dissection paradigm function together to strengthen the cases for frog dissection and against other options. This brings us to the first part of the materiality argument against frog dissection developed through this book: exposing and interrogating the assumptions of this paradigm of thinking. There is a legitimate debate to be had over whether we should prefer frog dissection or its alternatives. However, it is difficult to see these arguments for and against frog dissection with clarity when we are stuck within the context of the dissection paradigm, a context of preconceptions that make the frog dissection activity appear inherently preferable and its alternatives inherently inadequate.

What are the kinds of "intuitions" and "associations" and "preconceptions" at issue here? For one, there are questions of what should count as "real" in these discussions, as well as what the implications of such a designation should be for our understanding of something's educational value. A second thing at issue is whether an activity should be understood as akin to a "scientific experiment," and again what should be implied by such a status. Another central aspect of the intuitions and associations at issue in the dissection paradigm regard how we should compare the different options within the debate between frog dissection and its alternatives. That is, there are intuitions and associations here regarding the structure of our thinking, how things should be organized and contrasted. There are, for example, presumptions often made about what alternatives to frog dissection should be expected to do, and how they should be evaluated. And finally, we can see patterns in the ultimate structure of debate, i.e., how the disagreements over frog dissection tend to be cast, such as how each side of the debate tends to be pigeonholed.

In what follows in this chapter, we'll explore the details of all of these features of the dissection paradigm. The chapters that follow in this book will complicate and challenge these assumptions.

In describing these intuitions and associations and preconceptions as a "pattern of thinking" or as a "paradigm," I mean to emphasize the fact that these are general observations that do not hold equally or in the same way across all the places where this paradigm has purchase. That is, we can expect the details and the degrees of strength of these features to vary widely, perhaps very widely, from country to country, region to region, school district to school district, classroom to classroom. By referring to these features as patterns of thought, or as a paradigm, the point is specifically to provide a broad generalization. We should absolutely expect different teachers to teach differently.

In addition, the goal of this chapter is merely to provide a broad outline of these features of the dissection paradigm. While I will cite and footnote a number of pieces of journalism, surveys, academic studies, and other sources, it is not my intention to provide any kind of full account of the dissection paradigm, something that would be a large task given the variation between regions and classrooms mentioned above. The empirical project of proving the details and intensities of the dissection paradigm is only started here. These claims are open to further empirical confirmation or disconfirmation. And they are subject to evaluation in terms of the observations and impressions of those who live within places covered by the dissection paradigm. My expectation is that for people who, like me, were raised in areas that are under the dissection paradigm, the following observations will resonate.

The following account of the dissection paradigm thus serves as a kind of intuition test. By identifying and showcasing these alleged features of the dissection paradigm, we can then subject them to critical analysis. It is precisely the status of many of these features as unreflexive assumptions that provides much of their power. They exert a strong influence exactly because most people don't think too much about them. If we can find ways to undermine the assumptions of the dissection paradigm, then we can set the debate over frog dissection and its alternatives on a more proper footing.

2.1 Some Context

As someone who grew up in the United States, frog dissection was so normal and expected that it was shocking to learn that it—or any kind of precollege animal dissection—is not a common educational activity across the world. The practice of dissecting a frog in high school or middle school biology classes has been banned in multiple countries, from India to Israel, and is severely restricted in many others.[1] And yet, for those of us living in those areas of the planet encompassed by the dissection paradigm—which include, for example, the United States and Canada—frog dissection is entirely commonplace. It is a customary part of high school or middle school education. When this practice comes up in the news, it is routinely referred to as a "rite of passage."[2] That is, in those parts of the world where the paradigm holds, the frog dissection activity as an educational experience has an established place in culture, a mystique, a cachet.[3] It is a traditional part of growing up.

In her review of the literature, Jan Oakley reports that in North America, 78–94% of biology and science instructors use frog dissection in their teaching, and 88–95% of students report having participated in a dissection activity in at least one point during their education, which in the United States equals 10–12 million animals killed per year for these activities (2013a, p. 4). According to a 2022 survey conducted by the National Anti-Vivisection Society (NAVS), 85% of precollege biology

[1] See: Telegraph Correspondent. (2008). "Dissection Controversy in School." *The Telegraph India online*. Jan 14th, 2008. www.telegraphindia.com/jharkhand/dissection-controversy-in-school/cid/628680.

Watzman, Haim. (1999). "Dissection Banned in Israeli Schools." *Nature*. 402: 845.

[2] See, e.g., Huntsberry, Will. (2015). "Dissecting A Frog: A Middle School Rite Of Passage." *NPR. All Things Considered*. Feb. 25th, 2015. https://www.wnyc.org/story/dissecting-a-frog-a-middle-school-rite-of-passage/.

Shine, Nicole. (2014). "The Battle Over High School Animal Dissection." *Pacific Standard*. Oct. 15th, 2014. https://psmag.com/environment/battle-high-school-animal-dissection-92391/.

Lush, Tamara. (2019). "Fake Frogs in School Dissections Eliminate Gross Out Factor." *Associated Press*. Dec. 31st, 2019. https://apnews.com/article/tampa-florida-animal-rights-animals-frogs-2387cfcc3b431a52e2d718d7c1412420.

[3] Interestingly, as Lynette A. Hart and colleagues observe, "Although the practice of dissection continues in precollege classrooms, uses of animals in college and professional school laboratories have sharply declined across recent decades. Paradoxically perhaps, veterinary schools have led the way in creating instructional software and moving away from the traditional consumable uses of animals in instruction" (2008, p. 51).

teachers in the United States report using animal dissection in their classes.[4] The American Anti-Vivisection Society (AAVS) estimates the number of vertebrate animals, such as frogs, that are killed specifically for dissection for US high schools per year to total around six million every year.[5]

If you're not one of those who have participated in these kinds of learning experiences during your high school or middle school years, then here's what the typical experience involves. The frog body is there before you on the desk or lab table, lying on its back, probably on a tray or wax paper or something. Its smell is striking. You may be instructed to feel the smoothness of the frog's skin, or to observe the membranes that cover the frog's eyes and eardrums. Perhaps you are instructed to investigate the frog's mouth. If so, you'll pry it open with a metal probe or tweezers, and you may be instructed to use scissors to cut the jawbone on either side to get it open wider.

The first task is to secure the frog in place. One way this tends to be done is by placing the frog on a tray lined with a deep layer of hard wax or a foam or vinyl pad, and pushing metal pins through the frog's wrists and ankles. Next, you'll follow your teacher's instructions to make your first cuts. This can be accomplished with a scalpel or perhaps a pair of sharp scissors. Typically, the first incision will be a vertical one down the center line of the frog's chest, and then some horizontal ones across the chest and across the waist, making the shape of a capital letter I. Now the skin and muscle of the abdomen can be peeled open to reveal the organs inside. The organs from here can be cut and pulled out as you go through identifying them.

The kind of dissection experience students have today is largely similar to how it has been conducted in the past. Lynette A. Hart and colleagues in a largescale review of the history of classroom dissection observe that, "Despite major advances in teaching topics such as genomics, cell biology, and computer literacy, laboratory use of dissection remains almost unchanged in intermediate and secondary education. This status quo is reinforced by commercial firms that provide catalogs for conveniently ordering laboratory materials, including dissection specimens" (2008, p. 119). The cultural status of classroom dissection within the

[4] Reported here: https://navs.org/insight/animals-used-in-education/.

[5] Reported here: https://aavs.org/animals-science/how-animals-are-used/dissection/.

dissection paradigm as a kind of revered tradition may be related in some way to this unchanging quality.

This can be contrasted with the kinds of experiences you could expect if you were to take part in a dissection simulation. There are many computer simulations available, but they tend to begin with an image of a frog onscreen lying in a way that mimics the start of the classroom frog dissection activity: motionless and belly up. The image may be hand-drawn or computer-generated, or may be a photo. The program directs you to make your "cuts," which, depending on the program, you might perform by clicking the correct space onscreen, or perhaps dragging the mouse across the line of the desired incision, or possibly by moving your finger across the touchscreen interface. The interior of the virtual frog onscreen may be revealed in different ways, again depending on the program. Perhaps the screen advances to the next image, one of the frog now cut open. Perhaps the action of opening up of the frog body is animated. Perhaps the correct action onscreen has the effect of initiating a video clip. The computer simulations of frog dissection available exhibit a wide range of fidelity. Some aim for photorealism where possible, others for a more cartoony quality.[6]

While many aspects of frog dissection can be addressed through computer simulation, it seems straightforwardly true to say that they do not create the same experience. For example, Norm Friesen has

[6] Some examples of frog dissection simulations include:

- The Froggipedia App, by Embibe. See: https://www.embibe.com/exams/froggipedia/.
- The Frog Dissection App, by Emantras Inc. See: https://www.frogvirtualdissection.com/.
- Frog Dissection, by Gizmos. See: https://gizmos.explorelearning.com/find-gizmos/lesson-info?resourceId=1097.
- Dissection Simulator: Frog Edition, by VictoryXR. see: vxr.direct/dissection-simulator-frog-edition/
- Froguts, by BioElearning Co., a free and simple 25-year-old program: https://thesciencebank.org/pages/froguts?_pos=1&_sid=7f8604f41&_ss=r.
- 3D Frog Anatomy, by Biosphera. see: https://biosphera3d.com/product/3d-frog-anatomy-software/.
- 3D Dissectible Frog Model, part of the Visible Body Suite: https://www.visiblebody.com/blog/no-frog-no-prob-a-3d-visual-guide-to-frog-anatomy.

contrasted these experiences and notes the "visceral character" and "sensual assault" of frog dissection, and its relative absence in the case of computer simulations (2011, p. 90). On computer-simulated dissections, he writes,

> The manifold sense impressions that assaulted the body in the description of the in-school dissection are either absent or very much muted in this virtual activity: There is no smell of formaldehyde or rotting flesh; there is no need to fear that blood or any other liquid might come 'spurting' from the creature to be dissected... Handling and even cutting into the animal, furthermore, is a question of clicking on the correct virtual instrument (the scalpel) and gliding it, in effect, across part of the computer screen. (Friesen, 2011, p. 91f)

While it will be important to critically analyze the educational benefits or detriments that accompany the "visceral" nature of dissecting an animal in class, we can note here these clear differences in the raw experience of taking part in the frog dissection activity compared to that of a computerized simulation.

In addition to computer simulations of dissection, there is another important style of dissection alternative to consider: physical simulations. These tend to take the form of a frog-sized rubbery dissectable model. Students use a scalpel and scissors to slice open the physical frog model in a manner that directly resembles the task and bodily engagement of frog dissection. As with computer simulations, the level of fidelity of the different physical simulations varies. Some are strikingly realistic and could pass for a dead frog at a glance. The fidelity and complexity of the insides also vary. Some models are comparably very simple. Some offer the possibility of being sewn up again, so that they can be used multiple times. Of course, one advantage of physical dissection simulations is that they offer a kind of hands-on experience that is often touted as an important benefit of frog dissection.[7]

Another part of life within the dissection paradigm is the existence of a vocal minority who actively opposes the use of animal dissection in classrooms. Sometimes the push to reduce or eliminate practices of classroom dissection comes as part of the larger animal advocacy movement. There are many groups, non-profits, and other organizations across the world devoted specifically to raising awareness of the problems with animal dissection in schools. Many of the more influential advocacy organizations in this regard exist as a branch of a larger group. In particular, some activist groups that focus specifically on issues of ending animal testing and other harmful usages of animals in science and industry will also have an arm that advocates against animal dissection in schools, such as the National Anti-Vivisection Society (NAVS) based in London, Animal-Free Science Advocacy out of Melbourne, or the American Anti-Vivisection Society (AAVS) based in Pennsylvania.[8] These efforts are also taken up by animal advocacy groups more generally, such as organizations with a goal of reducing animal cruelty in its various forms, from factory farming to dog fighting. For example, organizations such as the Humane World for Animals (formerly the Humane Society of the United States and the Humane Society International), the Physicians Committee for Responsible Medicine, the Animal Welfare Institute, People for the Ethical Treatment of Animals (PETA), and others all have campaigns that address animal dissection in schools.[9] Similarly, defenders of animal dissection in

[7] Some examples of physical frog dissection simulations include:

- SynDaver's highly realistic SynFrog Dissection model:
  https://syndaver.com/product/synfrog-frog-dissection-model/.
- Kind Frog, a simpler dissectable frog model with color coded organs:
  https://www.kindfrog.org/.
- Other simpler toy models are available, e.g.,:
  https://topsecrettoys.us/products/dissect-it.

[8] For links to the websites of these organizations, see:
https://navs.org/.
https://aavs.org/.
https://animalfreescienceadvocacy.org.au/.

[9] For links to the websites of these organizations, see:
https://www.humaneworld.org/en
https://www.pcrm.org/news/good-science-digest/dissection-alternatives-students.

schools find support from larger institutions as well, such as associations of biology teachers looking to preserve teacher autonomy, scientific equipment suppliers aiming to preserve their market, and biomedical research organizations who defend animal experimentation in science generally and who approach educational dissection as a related issue.

A flashpoint in the history of criticism of classroom frog dissection is the work of Jenifer Graham. Back in 1987, Graham, then a 15-year-old high school student in California, refused to participate in a frog dissection activity, and when faced with a bad grade as a consequence of this refusal, took the school to court. This brought attention to the issue (including a CBS special and an Apple commercial), and led to the state passing a law that enables students to forgo the dissection activity and learn the same lessons another way.[10] Today, many countries or regions where the frog dissection activity is commonly practiced in schools will also have rules that enable students to "opt out" if they have reservations, such as moral or religious reasons for choosing to forgo participation. These are often referred to as "student choice" laws or initiatives. For example, in the U.S., 22 states currently maintain laws or some other state policy that upholds a student's right to refuse to engage in classroom dissection.[11] These laws often require such refusal to be backed by a letter from the student's parent or guardian. And they often call for schools to be ready to provide an adequate alternative activity. While these laws or policies provide rules for public schools, they do not usually put restrictions on private school practices.

The text of these laws tends to be direct and straightforward, if simply put. For example, the law in the state of Pennsylvania in the US reads: "Public or nonpublic school pupils from kindergarten through grade twelve may refuse to dissect, vivisect, incubate, capture or otherwise harm or destroy animals or any parts thereof as part of their course of instruction." It continues that such students, "shall be offered an alternative education project for the purpose of providing the pupil an avenue for obtaining the factual knowledge, information or experience required

https://www.peta.org/issues/animals-used-for-experimentation/animals-used-experimentation-factsheets/dissection-lessons-cruelty/. https://awionline.org/content/humane-education.

[10] For an account of the influence of Graham and others' work on educational policy, see (Orlans et al. 1998).

[11] See: https://aavs.org/animals-science/laws/student-choice-laws/.

by the course of study."[12] Sometimes, more guidance and context is provided, for example by noting things like educators' general approval of classroom dissection, and also the effectiveness of alternatives. The Massachusetts Department of Education has released a science and engineering curriculum framework which notes that the official policy is that: "All public schools that offer dissection as a learning activity should, upon written request by a student's parent or guardian, permit a student who chooses not to participate in dissection to demonstrate competency through an alternative method." It goes on to provide additional context, stating that, "Educators teaching life science consider dissection to be an important educational tool. But dissection should be used with care. When animal dissection is considered, teachers should recognize that there are other experiences (e.g., computer programs) for students who choose not to participate in actual dissections."[13] It is possible for these laws to go further. Illinois provides an interesting case. The text of the state's "Dissection Alternatives Act" first notes at length the huge importance of animal dissection for the history of biological research and human well-being (e.g., "Without dissection the science of anatomy could not have advanced, and it is the bedrock supporting the modern practice of surgery in its many forms"). It then goes on to note the significant advancement that has taken place for dissection alternatives over the years, going so far as to state that, "In certain circumstances these new technologies are capable of providing an educational experience superior to dissection, and they have often proven to be less expensive and more humane." It notes as well that, when inappropriately done or poorly supervised, dissection can be considered cruel. Its final recommendation is that schools in Illinois should: "make available and use alternatives to dissection when those alternatives are appropriate and can provide an educational experience that is equal or superior to the traditional use of dissection. It is not in any way the intention of this Act to discourage the

[12] For the text of the PA student choice law, see: https://www.palegis.us/statutes/unconsolidated/law-information/view-statute?SESSYR=1949&SESSIND=0&ACTNUM=14&SMTHLWIND=&CHPT=15&SCTN=23&SUBSCTN= .

[13] See pages 185-192 of the *2016 Massachusetts Science and Technology/Engineering (STE) Curriculum Framework* released by the Massachusetts Department of Elementary and Secondary Education.

https://www.doe.mass.edu/frameworks/scitech/2016-04.pdf.

appropriate use of dissection in research or when it provides a valuable educational experience to students."[14]

Studies show that biology teachers living within regions where these practices are common tend to strongly prefer the use of animal dissection in the classroom. For example, Pamela Osenkowski and colleagues recently surveyed more than 2,500 American biology teachers on the subject. They found that while the most would allow a student to do an alternative assignment if requested, the majority also "believe dissection to be the best way to teach anatomy and/or physiology" and that "the use of dissection in biology is an important tradition" (Osenkowski et al. 2022, p. 239). They also found that while the majority agreed that many of the core learning objectives can be met through the use of alternatives, most of these teachers state a preference for the animal dissection activity and believe that alternatives do not provide the same hands-on experience (Osenkowski 2022, p. 237). These survey results are in accord with earlier studies and surveys of other regions where precollege dissection is the norm.[15]

In the US context, we can look to the official statements released by biology teacher organizations for further confirmation of these views. For example, The National Association of Biology Teachers (NABT) official statement on "The Use of Animals in Biology Education" acknowledges that dissenting students should be provided with alternatives, but maintains that:

> While the increased quality and accessibility of dissection simulation software has helped address concerns from students and parents opposed to dissection, these alternatives are not without limitations. Utilizing a software-only approach may constitute a disservice to many students and does not acknowledge the well-documented educational benefits of hands-on dissection. (NABT, 2019)

[14] See: https://ilga.gov/Documents/legislation/ilcs/documents/010501120K5.htm See also: www.isbe.net/Documents/DissectionAlternatives-Fact-Sheet.pdf.

[15] For example, back in 2004 Lesley A. King and colleagues surveyed nearly 500 American high school biology teachers and found similar results. Similar results were also found in more recent surveys of high school teachers in Switzerland, Morocco, as well as Ontario, Canada (Zemanova 2022; Amahmid et al. 2019; Oakley 2012b).

The National Science Teaching Association (NSTA) statement on "Responsible Use of Live Animals and Dissection in the Classroom" similarly "opposes regulations or legislation that would eliminate an educator's decision-making role regarding dissection or would deny students the opportunity to learn through actual animal dissection" (NSTA 2008). They support the use of animal dissection activities since they can help students develop observational skills, learn about organisms' structures, and "develop a greater appreciation for the complexity of life" (NTSA 2008).

While biology teachers in general tend to express support for accommodating students who make the choice not to participate in the frog dissection activity, there are reasons to question whether this support actually always comes across to students. Oakley's survey work with teachers and students in Ontario, Canada, is suggestive. For example, she has found that when students objected to participating in dissection activities in school, only a third were provided with an alternative task (Oakley 2013b). The rest met with many different experiences, from being pressured to participate, to being asked to watch others do the activity, to being asked to find their own alternative task, to receiving a failing grade. Oakley has also interestingly found students' and teachers' own reporting on classroom experiences to differ. For example, another survey study she conducted in this region found teachers to underestimate the percentage of students who objected to the class dissection activity (with those teachers surveyed estimating 1–5% of students to be objectors, and, in contrast, 24% of students reporting objections), as well as to have offered alternatives at a higher rate than students report (with 73.7% of teachers surveyed claiming to offer the choice to opt out, and 60.8% of students claiming to have received such an offer) (2012a). These survey data additionally reveal a divergence in accounts of whether any discussion of the ethics of dissection took place in the classroom. Oakley reports that,

> There was also a considerable difference between the two groups' reported experiences with classroom discussions on the ethics of dissection: only 28.9% of the student sample reported that their teacher held a discussion on the ethics of dissection, whereas 86.3% of teachers reported conducting such a discussion as part of their practice. (Oakley 2012a, pp. 18–19)

It seems that in at least in some classrooms, students may not always feel fully free to opt out of dissection activities. If we truly want students to have a choice in this matter, then they should be able to exercise this choice in a way that does not require them to forcefully object and then stick to this objection in the face of pressure.

The central empirical question that emerges in discussions over educational dissection and its alternatives is: *Do alternatives teach as well?* The data on this are clear: *Yes.*

Studies on the effectiveness of dissection alternatives largely show them to produce at least the same learning outcomes. These studies have taken place over the last few decades, and they have compared dissection simulations to the hands-on dissection of many different animals, and they have investigated education at multiple levels, including college and professional training levels, and not only the precollege context on which this book focuses. Across these studies, the consensus is that simulated dissections teach equally well, or better. Or as Miriam A. Zemanova puts it, "there is a chasm between the evidence of the efficacy of humane teaching methods and the continued implementation of harmful animal use in education" (2022, p. 2).

For example, in an authoritative recent systematic review of 20 studies published between 2005 and 2020, Elizabeth Ormandy and colleagues found that "Our data show that in 95% of studies, students do at least as well—and in most cases better—when they use NAMs [non-animal teaching methods] compared to animal dissection" (2022, p. 402). These results accord with earlier systematic reviews of the literature as well (e.g., Zirkel and Zirkel 1997; Balcombe 2006; Patronek and Rauch 2007). So, it cannot be concluded that the strong teaching results of simulations only correspond to the advance of technology. For example, back in 2007 Gary J. Patronek and Annette Rauch conducted a systematic review of 17 studies that took place between 1996 and 2004. They report that, "In all 17 studies, results associated with the alternative method of instruction were either not significantly different from or superior to results associated with the conventional method of instruction" (Patronek and Rauch 2007, p. 38). We can even reach back to Jessica B. Zirkel and Perry A. Zirkel's review of seven empirical studies that took place through the 1990s, as well as a handful from the decades before (1997). They conclude that, "Although limited by several notable threats to both internal and external validity, the relevant research is relatively consistent across the various educational contexts, or levels, in terms of finding high-tech (e.g., IVD) [computer-based interactive videodisk] simulations for teaching physiological information to be as effective as actual dissection. At the same time, the quantitative analyses were consistent in terms of the low-tech (e.g., video) simulations similarly not yielding significantly different achievement outcomes from either their high-tech or live dissection counterparts" (Zirkel and Zirkel 1997, p. 53).[16]

These results resonate with the expansive and detailed recent systematic review by Zemanova and Andrew Knight which considered 50 relevant studies that have taken place between 1968 and 2020. Across this timeline, the humane teaching alternatives studied vary widely, with most evaluating computer simulations of dissection, but others considering the use of video, models, mannequins, or clay sculpting exercises. They found 90% of these studies to conclude that alternative teaching methods yielded equivalent or superior learning outcomes (with 60% finding an equivalence, 30% finding alternatives to yield superior outcomes, and 10% to yield inferior results). They conclude that these results, "overwhelmingly demonstrated that harmful animal use for teaching and training is not justified" (Zemanova and Knight 2021, p. 13).

2.2 Hands-on Learning and Individual Moral Preference

My suggestion is that under the dissection paradigm, debates over these issues tend to be couched within particular terms. Of course, there are many arguments that have been offered for and against frog dissection, and for and against its alternatives (and we'll review some of these in Sect. 2.6). However, I suggest that within the dissection paradigm, these

[16] To all of this we can add Jonathan Balcombe's detailed 2006 review of more than thirty studies that compare humane teaching methods (including computer simulations, but also the use of video tapes, 3D models, or even just lectures) to classroom animal dissection across the 1980s, 90s, and early aughts. As Balcombe concludes, "Collectively, the empirical evidence provides solid support for the replacement of traditional learning methods that involve harming and/or killing animals. Even in cases of equivalence in learning performance, ethics should dictate that the alternative learning method be used" (2006, p. 48).

One interesting study that bucks this trend was conducted by Tina R. Cross and Victor E. Cross in 2004. The learning outcome investigated in this small-sample experiment was how well simulation prepares students to do an exam involving the dissection of a dead frog, and vice versa. They find the frog dissection activity to do a better job than simulation in this regard.

disagreements tend to take shape in a more general manner. In the more typical framing, the two sides of the disagreement tend to be generalized as follows: On the one side, the frog dissection activity is boasted as a hands-on learning experience. On the other side, it is recognized that some students hold moral reservations about this activity that may call for accommodations. In this way, we receive a general setup within the dissection paradigm in which the frog dissection activity is understood to be valued for the educational experience it creates, and alternatives are valued for something else entirely, i.e., the way they provide an outlet for accommodating those who require different learning tools for personal reasons.

Perhaps the most salient difference between frog dissection and its alternatives—especially *computer-simulated* alternatives—is the issue of hands-on learning. This value of the frog dissection activity as a tactile and otherwise sensorily engaging experience is one of the most difficult things to reproduce in alternatives. Defenders of frog dissection readily point to this hands-on quality as a crucial feature. This point, as I will elaborate below, is simultaneously both one of the strongest and weakest of those on offer in defense of frog dissection.

We can see the centrality of this point—that frog dissection is valuable because it creates a hands-on experience—throughout these discussions. It is the primary reason that educators give for preferring the frog dissection activity over alternatives (King et al. 2004; Osenkowski et al. 2022). As one biology instruction website puts it, "frog dissections offer students the chance to engage in practical application of scientific principles. From proper dissection techniques to the identification of organs, students learn to follow procedural steps, make observations, and draw conclusions—skills that are fundamental to scientific inquiry. This hands-on approach not only reinforces theoretical knowledge but also cultivates critical thinking and problem-solving abilities."[17] Or, for example, a website for a science education equipment supplier notes that: "Many individuals find it much easier to learn when they can carry out physical activities. Indeed, kinaesthetic learning does boast greater benefits for some, enabling them to more fully engage with the subject matter being studied."[18] One defender of frog dissection puts the concern this way:

[17] See: https://www.biologycorner.com/myimages/frog-dissection/.

[18] See: https://vittaeducation.com/the-importance-of-dissection-in-biology/.

> The trend toward eliminating actual experiences for students, instead substituting some pseudoexperience is really troubling me. They aren't doing the dissection to learn that a pig has a gall bladder, or even that it is found in the underside of one lobe of the liver. They're doing the dissection so that they will have handled and seen the inside of a real animal in a disciplined way, and will have some basis in experience for the next time having had the experience might prove useful. Please reconsider abandoning dissections in favor of pictures, whether the pictures come packaged with a computer or not. What's already lacking in biology classes is hands-on experience.[19]

These kinds of comments reflect how much the frog dissection activity has become recognized for its value in creating a hands-on learning event, one that creates an experience that involves learning through doing.

This central point made in defense of frog dissection is both very strong in one sense, and yet extremely weak in another. It is strong in the sense that *we can all understand and appreciate the value of creating learning experiences in the classroom*, of educational activities, and of gaining knowledge through participation in actual undertakings. For example, someone who supports progressivism in education, who follows thinkers such as John Dewey, and who advocates for the importance of hands-on learning experiences that connect to students' lives could find the frog dissection activity to be in some ways strongly in tune with these values.[20] Personally, I find the idea quite persuasive that we should encourage the creation of vivid and relevant hands-on classroom experiences whenever possible. The frog dissection activity certainly has these qualities: it's striking, memorable, and it is interactive in a hands-on manner as students use tools to slice up and explore the frog body.

And yet, at the same time, *this argument is also quite weak because support for hands-on learning is not specific or exclusive to animal dissection.* That is, there are many good reasons to support hands-on learning activities in general, but these same reasons do not require us to support the dissection activity in particular. As Jonathan Balcombe notes, "The hands-on argument is not conclusive. Hands-on learning is not the exclusive domain of animal dissection. Many hands-on materials and approaches

[19] This comment is by biology professor Dave McNeely, made in back in 2000, and quoted in Cross and Cross (2004, p. 409).

[20] See (e.g., Dewey 1916; Vaughan 2018).

for learning animal biology do not require killing or harming animals" (2000, p. 7).

Put another way, someone could consistently argue two things: (1) in support for developing more and better hands-on educational activities for school students, AND (2) against dissecting animals in particular. Someone could consistently be a strong supporter of the principles of progressive education in the Deweyan sense mentioned above, and have reasons to believe that the dissection activity violates these same principles. (We'll return to Dewey and progressive education later in the book, but in criticism of dissection). Someone could consistently support hands-on learning in science (including the use of various scientific instruments from microscopes to test tubes), and also across the curriculum (including everything from map making to class trips), and at the same time be against the particular hands-on activity of frog dissection, whether it be for moral, environmental, or, as further argued throughout this book, educational reasons. Support—even strong support—for hands-on educational experiences does not necessitate support for frog dissection.

However, in my observation, this point about the separability of support for hands-on education in general and support for the frog dissection activity in particular is widely overlooked within the context of usual thinking within the dissection paradigm. It is the kind of point only made by active critics. That is, it is routine to see defenses of frog dissection made primarily or even exclusively in terms of the value of hands-on education, independent of whether these defenders themselves ever otherwise spend their time advocating for other hands-on educational experiences. Put into the language of implicit assumptions, we can say that: within the context of the dissection paradigm, challenges to the status quo tend to activate associations between this specific activity and a more general concern for hands-on learning.

I suggest that the dissection paradigm also generally includes an acknowledgment of the reservations that some students maintain about frog dissection. This recognition, as with all parts of the paradigm, represents only a trend or a tendency; we can certainly still find cases in which students are pressured or forced to do this activity. However, it can be noted that a part of the general discussion under the dissection paradigm can at times include some level of tolerance for student objections, but in a specific manner: they are taken as individual preferences to be accommodated.

We can see this interpretation of student objection throughout many institutions within the dissection paradigm. As noted above, associations of science teachers in the United States have released official position statements in support of the use of classroom frog dissection. These include the recommendation that objecting students should be accommodated. For example, the NSTA statement includes the recommendation that teachers: "Be prepared to present an alternative to dissection to students whose views or beliefs make this activity uncomfortable and difficult for them" (2008, p. 2). The NABT statement notes that, "teachers must be sensitive to the beliefs of each student and their right to make informed decisions concerning participation in dissection and, if possible, provide meaningful alternatives in keeping with course goals and objectives" (2019, p. 1). This kind of perspective is reflected in student choice laws, i.e., regulations which require that students have the option to forgo taking part in dissection activities.

It is a good thing that under the dissection paradigm students with objections are at least sometimes provided accommodation. It is a good thing when the relevant institutions such as state laws, school and classroom policies, and the official recommendations of teaching organizations support students' ability to opt out of participating in dissection. However, we should additionally add that this all also has the effect of further cementing the place of frog dissection as the status quo, the default, the expectation. And it has the effect of further casting the moral problems of the frog dissection activity merely as issues of personal preference. In this way, rather than understand moral criticisms as condemnations of the frog dissection activity and the industries that sustain it, they are understood as expressions of individual difference. Rather than understand moral criticisms of frog dissection as a form of advocacy, they are cast as the sensitivities of particular people that, at most, should be humored.

2.3 The Real and the Virtual

Another key feature of the dissection paradigm is the stipulation that the frog body presented to students is something real. And, correspondingly, the simulated alternatives—and especially computer-simulated ones—are understood to instead present students with something virtual. In a related manner, under the dissection paradigm the frog dissection activity is taken to present students with an encounter with a natural thing, and a simulation, instead, an artificial one.

This is of course quite reasonable, at least on its face. The frog lying on the table before the student is a physical thing there in the world. So are the various conventional tools of the activity, such as the scalpel, pins, and forceps. These physical things are taken up in a hands-on manner, one that engages the students' bodily senses, with the stiff and papery flesh of the frog providing a kind of physical resistance to snips and slices, and resulting in notable sights and smells. What's more, the frog body had grown in the world as a natural object. The frog body, there waiting to be dissected, had a life beforehand where it had hatched from an egg and matured and lived. It has not been fully made by humans, bit by bit, like an artist would make a clay sculpture, or a watchmaker would set the gears of a watch into place. This is another way in which the frog of the dissection activity is something we think of as a "real" thing.

Simulated alternatives instead involve virtual things. Again, this seems like a reasonable characterization, at least on its face. Both physical dissection simulations and computer-simulated dissection programs are artificial things that have been made by people. This human-made quality is one sense in which these simulations can be referred to as virtual things. And in the case of computer simulations in particular, the entirety of the content occurs within a virtual space onscreen, with a digital version of a frog body made available to students by programmers for a specific kind of observation and interaction. The entirety of the learning experience has been pre-programmed, at least in some significant sense. All of these kinds of things are being pointed to when dissection simulation is referred to as "virtual."

However, I suggest that there are additional implications that are also smuggled in by these designations, at least in their operation within the context of the dissection paradigm. That is, while it may seem obvious to refer to frog dissection as an activity that provides students with an experience of something real, and simulations an experience of something virtual, this characterization *also* often carries in hidden value judgments, and covers over other relevant aspects of these activities, all of which should be subject to critical reflection.

One observation to make about these designations of the real and the virtual is that they may not always be used merely as innocent descriptors. While the terms "real" and "virtual" seem to straightforwardly apply to the frog body and the simulated frog, respectively, the act of using these terms may often have normative implications. That is, the terms real and virtual, when used within the context of the dissection paradigm, may often be made as part of an implicit value judgment. I suggest that within the dissection paradigm, there is an implicit hierarchy understood to exist between the real and the virtual. Within the context of the dissection paradigm, the act of referring to a simulated experience as something virtual can carry with it a negative connotation. At least sometimes when these terms are used, it could be as part of a move made by defenders of frog dissection to cast simulation as providing an experience that is *merely* virtual.

I suggest that this binary designation in which one side of this debate is constituted by real things, and the other side by virtual things, can have another obscuring effect: it can downplay the human-made aspects of the dissection activity. That is, in labeling the frog dissection activity as something real, and in contrasting it with an entirely and unequivocally human-made thing like a simulation, this common setup of the dissection paradigm can have an effect of drawing our attention away from the fact that the frog body awaiting dissection is not a simple found object. The frog dissection activity does not simply and exclusively constitute any kind of direct experience of the natural world. This activity has also been significantly designed and curated. The objects have been pre-prepared. These curations and pre-preparations are educationally and morally relevant. And while there are multiple pre-prepared and curated aspects of the frog dissection activity that we will explore in the following chapters, the central morally and educationally relevant element of curation and pre-preparation is surely already obvious: the frog-for-dissection has been killed.

2.4 The Ideal and the Copy

I suggest that another central feature of the dissection paradigm is the idea that the main goal of alternatives to frog dissection should be to reproduce the experience of the frog dissection activity as closely as possible. That is, under the dissection paradigm, alternatives are assigned the role of mimicry. This is at once both entirely understandable, but also, I suggest, a situation which has the effect of reinforcing many of the core troublesome biases of the dissection paradigm.

Computer simulations of dissection routinely take the form of an echo of the dissection activity. The simulated frog appears onscreen, laid out, lifeless, tummy up, in the same position as the physical frog would lie before a student on the table in the classroom. With a hand on the mouse or a finger on the touchscreen, the student performs a pantomime of "slicing" the virtual frog. The physical simulations are similar, again with the simulated body splayed in the position of the frog awaiting dissection. The interface is similar: scalpels and cutting tools.

Again, this is perfectly understandable: the objective of this kind of direct mimicry is to show that the simulation can serve as an adequate stand-in for the frog dissection activity. There is a straightforward element of animal advocacy here: if these simulations can pass themselves off as good enough and similar enough educational experiences, then an argument is made for swapping them in since they avoid the fraught ethical issues attached to the traditional frog dissection activity. For proponents of dissection simulations, this kind of argument is seen as especially persuasive when paired with the moral and environmental concerns raised about the frog dissection activity, as well as with the empirical studies that largely show simulations to be at least as educationally effective.

However, I also suggest that all of this has an effect of reinscribing a core aspect of the dissection paradigm: that the frog dissection activity itself is the educational ideal in this scenario, and that, accordingly, the aim of alternatives should be to reproduce the same experience as closely as possible.

As you might expect, this setup puts dissection at a distinct disadvantage. The act of dissecting a frog body constitutes a very specific and very vivid experience. There are many parts of this experience that are difficult, if not impossible, to capture through an alternative format, and especially one set within a computer screen. And insofar as the frog dissection activity is implicitly regarded as the ideal in this scenario, then any differences that can be identified are thus automatically cast as deficits.

This setup also has the effect of directing attention away from the critical analysis of the frog dissection activity itself. When cast as an educational ideal, as the standard of comparison, as the very thing that any alternative can only ever hope to come close to replicating, then not much room is left for stepping back and evaluating the frog dissection activity in its own terms. That is, not much room is left for doing an assessment of frog dissection's *own* educational advantages and disadvantages. Simulation too, I suggest, is hindered by these expectations of the dissection paradigm; rather than consider the distinctive educational benefits that could result from the exploration of these same and related lessons through alternative means, simulations are stuck with a primary task of mimicry.

Put another way, the dissection paradigm directs attention away from a critical analysis of the frog dissection activity and its alternatives, and instead refocuses discussion mainly on issues of simulation fidelity.

2.5 Dissection as Laboratory Experimentation

I suggest that another aspect of the dissection paradigm is a tendency for the frog dissection activity itself to be thought of as a kind of scientific laboratory experiment. By participating in this classroom learning activity, students are understood to have the opportunity to "do science," in at least some limited sense. Students are of course not testing new hypotheses or generating new knowledge, but merely taking part in a pre-planned educational task. However, students do investigate specimen in a disciplined and careful manner (or at least hopefully they do), and they may make discoveries that are, at least, new to them. The dissection activity, like similar learning activities in other science classes, is often referred to as a "lab." Students often fill out "lab reports."

This conception of the frog dissection activity as a kind of quasi-experiment is also reflected in the concerns of both defenders and critics of the practice. Defenders of frog dissection sometimes lament that were it to be replaced by alternatives, future scientists and doctors would be denied an early instance of hands-on training with an actual animal specimen, an experience that helps students recognize whether such a life is right for them.[21] Others have alleged the opposite, i.e., that such an early experience of visceral and perhaps moral discomfort may have the effect of turning some students away from potential future careers in science and medicine.[22] Either way, these kinds of comments reflect the ways in

[21] For example, Terry D. Keiser and Roger W. Hamm argue that some students "may eventually enter a medical (human or animal), medical technology, pharmacy, or health-related profession. Dissections will provide these students the with the knowledge necessary to be competitive when preparing to enter these fields" (1991, p. 15).

[22] For example, Omar Amahmid and colleagues review the state of educational frog dissection in Morocco and find it to be commonplace, with 83% of teachers surveyed, as well as 61% of students viewing it as a positive experience (2019). For the not insubstantial proportion of students who felt otherwise, these authors suspect that, "These negative emotions may weaken students' interest in Life Sciences, and influence negatively on their cognitive performance and learning achievement" (Amahmid et al. 2019, p. 77).

which under the dissection paradigm the dissection activity is associated with a notion of actual science participation.

My suggestion is that this conception of frog dissection is not innocent. Insofar as it is true that under the dissection paradigm the dissection activity is conceived to be akin to an experiment conducted in a science lab, there are a number of consequences to keep in mind. Let's consider three.

First, we can note that if the frog dissection activity is understood as an example of students doing science, then this brings along certain connotations. The practice of science is associated with many values. For example, science is both commonly and professionally understood to be associated with "objectivity." Of course, philosophers of science and other analysts of scientific practice recognize objectivity to be a complex and multifaceted concept.[23] However, with regard to the patterns of assumptions and preconceptions of the dissection paradigm, we should consider how the understanding of classroom frog dissection as an instance of actually doing science may carry in associations with a naïve conception of objectivity. If the frog dissection activity is implicitly associated with a naïve understanding of objectivity and scientific truth, then this may also by implication make it seem that things such as moral concerns and other judgments have no place here. If, under this kind of implicit reasoning frog dissection is associated with things like science and laboratories and objectivity and facts, then any moral concerns may by implication be accordingly categorized as merely subjective things, things that perhaps pertain only to a student's individual feelings and opinions. In this way, a conception of the frog dissection activity as a form of science participation may by implication function to distance it from the moral concerns raised by critics. Sure, such an implicit argument for distancing dissection from moral concerns appears specious. But since under the dissection paradigm it is made at the level of implicit implications and preconceptions, it may be persuasive to some, nevertheless.

[23] See, for example, Lorraine Daston and Peter Galison's excellent historical account of how the notion of objectivity in science has changed over time, and continues to do so (2007). For a useful overview of philosophical analyses of the notion of scientific objectivity, see: Reiss, Julian, and Jan Sprenger. (2020). "Scientific Objectivity." Entry in Edward N. Zalta (Ed.), *The Stanford Encyclopedia of Philosophy*.
https://plato.stanford.edu/archives/win2020/entries/scientific-objectivity.

Second, if the frog dissection activity is thought of as a kind of actual scientific experiment, then this gives dissection an advantage in the comparison against alternatives. Insofar as frog dissection is taken to be not only an activity for teaching anatomy, but also an opportunity for students to actually do science, any alternative has more to live up to. That is, this conception of frog dissection implicitly expands what kind of education it purportedly provides. Under this conception of frog-dissection-as-science-experiment, this activity implicitly supplies *more than* particular learning outcomes (which simulated alternatives appear to do at least equally well). It also implicitly supplies *more than* a hands-on learning experience (of which there are many possible options besides dissection, if perhaps none more iconic). Under this conception, frog dissection implicitly opens a door for students into the experience of doing science like a scientist. If the frog dissection activity is thought of this way, then alternatives are pressured to somehow also supply this special opportunity.

Third, we can think about what exactly the frog dissection activity teaches students *about doing science*. If frog dissection is couched as an early opportunity for students to do science, then what does this experience teach them about science itself and about what it means to be a scientist? Surely many things. However, one aspect of the experience jumps out as interesting in these terms: the revulsion many students initially have, and their work to compartmentalize, overcome, or persevere through this discomfort. For many students, a key part of the dissection experience, and exactly part of what makes it so memorable, is a feeling of revulsion (which may include both queasiness and moral unease), as well as the work of getting over those feelings. As Jonathan Balcombe notes, "Some science teachers even admit that one of their aims is to desensitize students" (2000, p. 18). Oakley, in reference to her studies cited above on student attitudes toward dissection, observes that, "There's this old-school thinking [about science] that you've got to man up and not show your emotions... And when I talk to students in my work, some of them had received that very message from their teachers: Suck it up."[24] What is taught to students about science through this experience of desensitization?

Since under the dissection paradigm the frog dissection activity is often understood as an opportunity to do science, these conspicuous experiences of disgust and desensitization may imply to students that these things are somehow an important part of what it means to be a scientist. Through hands-on training, this activity teaches a student that scientists approach animals as instruments. This activity instructs by implication an instrumental approach to animal life, and it further implies that this kind of instrumentalization is a part of science. Dorian Solot and Arnold Arluke have conducted anthropological research into classroom experiences with frog dissection on exactly this issue. On their observations, they write,

> Although dissection does not completely socialize future scientists, it is indisputably a site of indoctrination of Western values about science and animals. As such, dissection rites of passage, such as middle school biology class, are likely to exert some influence on the consciousness and expectations of those who go on to use animals in their higher education and future careers. (Solot and Arluke 1997, p. 50)

Solot and Arluke find that students respond to this exercise in desensitization in many ways, such as by shifting from thinking of the specimens as animals to instead approaching them as component parts, by defusing the emotional situation with humor, by focusing on positive aspects of the experience, and by taking up a "macho" posture and compartmentalizing their negative feelings. These coping mechanisms themselves can be understood as a kind of early science training. Or, echoing these kinds of observations of some biology classrooms, Steve F. Sapontzis argues that,

> Some macho biology teachers apparently consider overcoming a distaste for killing animals or for handling and cutting into dead bodies to be a goal of biology courses, one which cannot be accomplished without dissections... For example, some may believe that dissection is essential to a crucial mission of biology classes, namely, to maintaining the "integrity" of the biology profession by keeping out the squeamish, the sentimental, the

[24] Quoted in Averett, Nancy. (2020). "High School Dissections Are a Science Class Tradition. But Are They Doing More Harm Than Good?" *Discover*. Jan 20th, 2020. https://www.discovermagazine.com/the-sciences/the-argument-against-high-school-animal-dissections.

> effeminate, and "animal libbers." However, such a mission would clearly be morally objectionable and, consequently, could not provide a morally justifiable basis for refusing to make accommodations for the moral beliefs of students opposed to dissection. (1995, p. 183)

If the dissection activity trains students to see the work of scientists as involving a desensitized stance toward animal life, then this comes to the detriment of both the students and to science itself.

2.6 The Dissection Debate

Activists and academics have over the years offered some more nuanced arguments for and against animal dissection in schools. This is what may come to mind for some when they hear a reference to the debate over dissection. Like gun control or capital punishment, the academic debate over the ethics of classroom dissection is something that makes for the kind of topic about which teachers might assign students to write an essay. It is not uncommon to see outlines of the arguments "for" and "against" classroom animal dissection.[25]

I have been arguing in this chapter that much of the debate over educational dissection as it actually occurs in classrooms and public discourse tends to operate within a context of assumptions that we've been referring to here as the dissection paradigm. However, it is also useful to look at some of the explicit arguments offered by those engaged in direct debate on this topic.

Defenses of frog dissection include the emphasis of its importance for teaching lessons on anatomy and physiology, the highlighting of the limitations of alternatives, and the deflation of criticism.

As discussed above, the central defense of the frog dissection activity is that it provides a hands-on experience. This is touted as a main feature of this activity that distinguishes it from computer-simulated alternatives. As one biology professor nicely puts the point, "if you're not doing the dissection but, let's say, looking at a picture of it or a video of it—it's the difference of looking at a sculpture versus actually being able to touch

[25] See, e.g., the *Encyclopedia Britannica* online entry for "Animal Dissection: Should Students Dissect Animals?".

https://www.britannica.com/procon/animal-dissection-debate.

the sculpture."[26] In addition to this core argument about the hands-on element of the dissection activity, further arguments are also made. For example, since classroom dissection constitutes a striking and visceral experience, defenders claim that this makes the lessons taught all the more memorable. As Hart and colleagues write,

> With dissection, part of the experience is the smell of formalin and the squishy-looking organs in the abdomen. Students may feel a mixed approach-avoidance polarity in facing this experience; it is unpleasant and very intriguing at the same time. The presence of disgust plays a role in making the experience much more memorable, perhaps accounting for so many people being able to recall certain details from their dissecting experiences in high school, whereas most of their other classwork is long forgotten. (2008, pp. 137–138)

Defenders also point to the complexity of classroom dissection experience. These defenses often point out that the lessons imparted by the animal dissection activity should not be understood to reduce to merely the rote identification of internal organs; the dissection activity enables students to follow through the connections of the organs and structures within the body. As Terry D. Keiser and Roger W. Hamm observe, "Dissection is a hands-on experience that lets students actually look inside a member of the animal kingdom… The more one has to involve a variety of our sensory abilities in exploring an animal's anatomy, the greater will be our appreciation and understanding of its structure and function" (1991, p. 14). Or as Douglas Allchin, one of dissection's most eloquent defenders, puts it,

> a real body is messy. It is held together with all sorts of connective tissue. Its compartments are lined by layers of membranes. There's fascia and fat. The complexity just doesn't show up in textbooks—or dissection programs that (ostensibly serving education) make things simple. Demonstrating the gap between idealized textbook diagrams and reality is one extraordinary value of looking inside real organisms. (2005, p. 370)

[26] Quoted in: Bok, Oliver. (2015). "Professors Discuss Pros, Cons of Dissection in Intro Labs." *The Oberlin Review*. Apr. 27th, 2015. https://oberlinreview.org/7988/news/professors-discuss-pros-cons-of-dissection-in-intro-labs/.

It is on this issue where defenders of animal dissection emphasize the stark comparison with simulated alternatives. Simulations rarely capture this kind of detail. Instead, simulations tend to be comparatively seamless. As Norm Friesen notes, "The simulation—like any other computer interface—is designed in such a way that the student experiences minimal disruption, despite the complexities and discontinuities that underlie it" (2011, p. 105).

In addition, defenders of classroom dissection level counterarguments against the concerns raised by critics. A. David Kline offers perhaps the most thorough set of rebuttals, deflating ecological concerns, and offering a measured consideration of ethical and educational issues (1995). Resonating with points made by national biology teaching organizations mentioned above, Kline frames the issue in terms of the academic freedom of biology teachers themselves. This leads him toward a compromise position similar to what is common within the dissection paradigm, in which teachers should retain the autonomy to teach as they see fit, but students are also provided with the option of doing alternatives. However, we can note that attempts at dismissing the moral concerns of critics can at times veer into the sneering. For example, in a 1988 editorial in the journal *Science*, Daniel E. Koshland Jr. writes,

> Consider, for example, the moral shock of the young student who finds that the stomach of a real frog contains mosquitos, flies, and small grasshoppers. Far from being the beloved and harmless frog that she imagined, she finds a predator actually eating other species with no regard whatsoever for their rights. Letting that frog live condemns many mosquitos, flies, and other insects to their deaths. This moral trauma is inappropriate for an immature student who may then conclude that the world is not nearly as simple as she had imagined. A computer simulation could replace the stomach contents with materials such as potato chips, soda, and other emotionally neutral nutrients. (449)

The satire goes on to suggest that legislation should be enacted to anesthetize flies before they're swatted.

On the other side of the debate, criticisms of frog dissection include moral and environmental concerns, educational ones, and arguments in favor of alternatives.

As noted, the central argument against the use of animal dissection in classrooms is an ethical one. Critics argue that it is wrong for animals to be killed for an introductory-level educational activity, and one for which there are sufficient alternatives. For example, as the website for the American Anti-Vivisection Society reads, "Every year, millions of animals are dissected in elementary, middle, and high schools, with more used in colleges and universities. Most of these animals suffer and die for a one-time educational demonstration."[27] Arguments for the value of animal life come with different justifications, from arguments that weigh animal suffering against a classroom exercise, to assertions of animal rights. Critics also point out that the moral wrongs of the dissection activity consist not only of the animal's death, but also the suffering caused in the process of its capture and preparation. For example, while the exact details of animal sourcing and processing are often kept as trade secrets by suppliers, one important study by Erich Gibbs and colleagues from the 1970s outlines the pain and hardship that frogs endure during capture, but before they are killed. They report that, "At the present stage of the art, nearly every step of catching and shipping places severe stress on the frog" (Gibbs et al. 1971, p. 1029). As F. Barbara Orlans and colleagues explain,

> Harms can also come from the trauma of capture and transport of wild frogs to the biological supply house. The commercial trade sometimes holds the frogs for several days in sacks filled with other frogs. There they may suffer dehydration, contract disease, or die before arriving at the biological supply house. Whether the animal is alive, sick, or dead is not a financial incentive to different treatment, because the economic value of the animal is not affected. (1998, p. 215)

While students are presented with specimen ready for the learning activity, the process that has occurred to make that object of study available to them is one that has involved the animal's suffering and death. Critics go so far as to argue that the frog dissection activity is unethical not just for the harm it causes to animals, but for the unethical lessons it imparts to students. For example, Sapontzis argues that,

> The widespread use of such exercises requires the killing of millions of animals annually. If these exercises are conducted even though they are inessential to meeting the educational goals of biology classes, we are teaching students that animals are dispensable. Such inessential exercises

[27] See: https://aavs.org/our-work/humane-education/.

> teach students that animals can be killed for trivial purposes, for example, just for curiosity or just because it has become traditional to kill animals on these occasions. (1995, p. 185)

This final ethical point—that dissection teaches students by implication that animal life has little value—is relevant to this book's project of analyzing the educational strengths and drawbacks of the dissection activity and its alternatives. We'll return to this in detail later.

There are also environmental arguments against frog dissection. Since the frogs used in classroom dissection activities are largely caught from the wild, critics allege that the mass removal of these creatures has negative effects on the ecosystem. As a critical write up in *E-The Environmental Magazine* explains, "Since biological supply companies have been unable to breed frogs in a cost-effective manner, they rely upon an international network of harvesters and distributors for their wares... The massive collecting of frogs has depleted many local populations, leading some states—and some countries—to outlaw their commercial harvesting... The environmental cost includes disruption of the local food chain, too" (2004). Critics also warn that formaldehyde and other chemicals used to preserve animal specimen for dissection can both present health risks to children, and represent a toxic pollutant for the environment.[28] In this way, the criticism of the frog dissection activity consists not only of the harm done to the particular animal that was killed, but also the larger environmental consequences of this practice.

Of course, criticisms of animal dissection are also often strengthened by pointing out the availability and suitability of alternatives. Critics argue that if the frog dissection activity raises ethical and environmental concerns, then it should be replaced with other activities. These arguments are bolstered by the large body of empirical studies, discussed above in Sect. 2.1, that show alternatives to teach the same information as well, or better, than animal dissection (e.g., Zemanova and Knight 2021; Ormandy et al. 2022). As computer and physical simulations continue to advance, teachers and activists continue to outline their educational potential (e.g., Jukes and Chiuia 2006).

Proponents of alternatives to animal dissection point out additional advantages. A major one is cost. For example, as the NAVS website puts

[28] See, e.g.: AnimalLearn's Environmental Impact of Dissection factsheet, https://www.animalearn.org/pdf/2024_Environmental-Impact-Flyer.pdf.

it, "Education can become more inclusive, cost-efficient, and beneficial to all through humane methods that eliminate dissection and open up discussions about animal welfare."[29] A frog cadaver used in the dissection exercise can only be cut open once. A computer simulation can be used over and over, and this presents a potentially large cost savings depending on the circumstances. Physical simulations vary with regard to whether they cost more or less than frog cadavers, and whether they are reusable. Another benefit of computer simulations and reusable physical simulations is that students can potentially repeat the activity multiple times, learning through repetition. If one wanted to learn through repetition with conventional frog dissection, then this would at least require the use of multiple frog cadavers.

Balcombe summarizes the arguments against frog dissection this way:

> When one considers the associated costs—animal suffering and death in the supply trade, disruption of wild animal populations, messages that tend to undermine rather than reinforce respect for life and concern for others, rising costs of animal carcasses (as compared with alternatives with longer shelf lives), exposure to potentially harmful chemicals, and greater time expenditures in preparing and presenting various animal-based exercises—the balance clearly falls on the side of abandoning dissection, at least in its current form. (2000, p. 79)

For the critics of animal dissection, the moral and environmental arguments against dissection, along with both the teaching effectiveness and cost effectiveness of alternatives, add up together to make a convincing case to discontinue this classroom practice.

While technical arguments are offered for and against the usage of frog dissection in the classroom, my suggestion is that a stronger influence on our thinking on this issue comes from a set of background preconceptions and assumptions related to these practices. These backgrounded ideas concern things such as what counts as real and what counts as artificial in these discussions, whether the dissection activity should be understood as a science experiment, what the proper goal of dissection simulations should be, and more. And I suggest that these ideas function together to bias many people in favor of the frog dissection activity and against

[29] See: https://navs.org/ending-classroom-dissection/.

simulated alternatives. These background ideas—the dissection paradigm, as I've been referring to them—are powerful exactly because they exert their influence in the form of unreflexitive intuitions and preconceived notions, things taken for granted and seemingly barely up for dispute.

Since the dissection paradigm maintains such a strong hold over some places, within those spaces it can feel like an inevitability. For those living within the dissection paradigm, and especially for those who haven't thought about it critically, the usage of frog dissection in school biology can seem obvious, like common sense, like just good science. The paradigm constrains our thinking.

What will it take to challenge the hold of the dissection paradigm? In this chapter, we have begun to identify the paradigm for what it is. Next, we'll explore some ideas that can help highlight its contingency.

References

Allchin, Douglas. 2005. "'Hands-Off Dissection?' What Do We Seek In Alternatives to Examining Real Organisms?" *The American Biology Teacher* 67 (6): 369–374.

Amahmid, Omar, Youssef El Guamri, Mohamed Yazidi, Bouchra Razoki, Khadija Kaid Rassou, Youness Rakibi, Ibrahim Amine Farouk, and Fadya Charkaoui. 2019. "Animal use in Life Sciences Education: Current Status, Teachers' and Adolescents' Attitudes and Alternatives." *Anatolian Journal of Education* 4 (2): 69–80.

Balcombe, Jonathan. 2000. *The Use of Animals in Higher Education: Problems, Alternatives, and Recommendations*. Washington DC: Human Society Press.

Balcombe, Jonathan. (2006). "Assessment of Alternatives in Education." In *From Guinea Pig to Computer Mouse: Alternative Methods for a Progressive, Humane Education*, ed. N. Jukes and M. Chiuia, 2nd edn., 40–52. Leicester: InterNICHE.

Cross, Tina R., and Victor E. Cross. 2004. "Scalpel or Mouse?: A Statistical Comparison of Real and Virtual Frog Dissections." *The American Biology Teacher* 66 (6): 409–411.

Daston, Lorraine, and Peter Galison. 2007. *Objectivity*. New York: Zone Books.

Dewey, John. 1916. *Democracy and Education*. New York: Macmillian.

E Magazine Editors. (2004). "Harvest of Shame." *E-The Environmental Magazine*, July 20. https://emagazine.com/harvest-of-shame/.

Friesen, Norm. 2011. *The Place of the Classroom and the Space of the Screen: Relational Pedagogy and Internet Technology*. New York: Peter Lang.

Gibbs, Erich L., Georgie W. Nace, and Marvin B. Emmons. 1971. " The Live Frog is Almost Dead." *BioScience* 21 (20): 1027–1034.

Hart, Lynette A., Mary W. Wood, and Benjamin L. Hart. 2008. *Why Dissection? Animal Use In Education*. Westport: Greenwood Press.

Jukes, Nick, and Mihnea Chiuia. 2006. *From Guinea Pig to Computer Mouse: Alternative Methods for a Progressive, Humane Eudcation*, 2nd ed. Leicester: InterNICHE.

Keiser, Terry D., and Roger W. Hamm. 1991. "Dissection: The Case For." *The Science Teacher* 51 (1): 13–15.

King, Lesley A., Cheryl L. Ross, Martin L. Stephens, and Andrew N. Rowan. 2004. "Biology Teachers' Attitudes to Dissection and Alternatives." *Alternatives to Laboratory Animals* 32 (1) suppl: 475–484.

Kline, A. David. 1995. "We Should Allow the Dissection of Animals." *Journal of Environmental and Agricultural Ethics* 8 (2): 190–197.

Koshland, Daniel E. 1988. "Homo Photosyntheticus." *Science* 239 (4839): 449.

National Association of Biology Teachers. 2019. "The Use of Animals in Biology Education." Position Statement. Revised and Adopted by the NSTA Board of Directors, July 2019. https://nabt.org/Position-Statements-The-Use-of-Animals-in-Biology-Education

National Science Teaching Association. 2008. "Responsible Use of Live Animals and Dissection in the Science Classroom." Position Statement. Adopted by the NTSA Board of Directors, June 2005. Revised 2008. https://www.nsta.org/nstas-official-positions/responsible-use-live-animals-and-dissection-science-classroom

Oakley, Jan. 2012a. "Dissection and Choice in the Science Classroom: Student Experiences, Teacher Responses, and a Critical Analysis of the Right to Refuse." *Journal of Teaching and Learning* 8 (2): 15–29.

Oakley, Jan. 2012b. "Science Teachers and the Dissection Debate: Perspectives on Animal Dissection and Alternatives." *International Journal of Environmental & Science Education* 7 (2): 253–267.

Oakley, Jan. 2013a. *Animal Dissection in Schools: Life Lessons, Alternatives and Humane Education*. [Policy paper.]. Ann Arbor, MI: Animals and Society Institute. Available online at: www.animalsandsociety.org/research/digital-archive/.

Oakley, Jan. 2013b. "'I Didn't Feel Right About Animal Dissection': Dissection Objectors Share Their Science Class Experiences." *Society & Animals* 21: 360–378.

Orlans, F. Barbara. 1988. "Debating Dissection." *The Science Teacher* 55 (8): 36–40.

Orlans, F. Barbara, Tom L. Beauchamp, Rebecca Dresser, David B. Morton and John P. Gluck. 1998. "Dissection of Frogs: The Jennifer Graham Case." Chapter 11 of *The Human Use of Animals: Case Studies in Ethical Choice*. Oxford: Oxford University Press.

Ormandy, Elizabeth, Janella C. Schwab, Samantha Suiter, Nicole Green, Janice Oakley, Pamela Osenkowski and Christine Sumner. (2022). "Animal Dissection vs. Non-Animal Teaching Methods: A Systematic Review of Pedagogical Value." *The American Biology Teacher* 84 (7): 399–404.

Osenkowski, Pamela, Ignas Karaliunas, and Merari Diorio. 2022. " Educators' Views on the Use of Dissection Alternatives in American Biology Classrooms." *Alternatives to Laboratory Animals* 50 (3): 235–243.

Patronek, Gary J., and Annette Rauch. 2007. "Systematic Review of Comparative Studies Examining Alternatives to the Harmful Use of Animals in Biomedical Education." *Journal of the American Veterinary Medicine Association* 230 (1): 37–43.

Sapontzis, Steve F. 1995. "We Should Not Allow Dissection of Animals." *Journal of Agricultural and Environmental Ethics* 8 (2): 181–189.

Solot, Dorian, and Arnold Arluke. 1997. "Learning the Scientist's Role: Animal Dissection in Middle School." *Journal of Contemporary Ethnography* 26 (1): 28–54.

Vaughan, Kelly. 2018. "Progressive Education and Racial Justice: Examining the Work of John Dewey." *Education and Culture* 34 (2): 39–68.

Zemanova, Miriam A. 2022. "Attitudes Toward Animal Dissection and Animal-Free Alternatives Among High School Biology Teachers in Switzerland." *Frontiers in Education* 7: 892713.

Zemanova, Miriam A., and Andrew Knight. 2021. "The Educational Efficacy of Humane Teaching Methods: A Systematic Review of the Evidence." *Animals* 11: 114.

Zirkel, Jessica B., and Perry A. Zirkel. 1997. "Technological Alternatives to Actual Dissection in Anatomy Instruction: A Review of the Research." *Educational Technology* 37 (6): 52–56.

CHAPTER 3

The Technology of the Dead Frog

Abstract In discussions on the use of animal dissection in precollege classrooms, many assumptions are often made which favor the dissection activity over alternatives. This chapter explores the potential for using ideas from the fields of philosophy of technology and STS to expose and undermine such presumptions. In particular, these ideas can be used to highlight the statuses of both animal dissection and its simulated alternatives as concrete material practices. Rather than conceive of animals used for classroom dissection as "real" or "natural" specimen, we can also emphasize their situation within this activity as specific classroom instruments, removed from their natural habitat, and killed and prepared for usage. This reconceptualization sets up further explorations of their specific educational advantages and limitations.

Keywords Frog dissection · Educational technologies · Biology education · Simulation · Animal welfare · Humane education · Technological mediation · Philosophy of technology · Postphenomenology

Discussions on frog dissection always note how unforgettable and hard hitting this activity can be. I remember my own high school experience of it. However, one of the things I recall most clearly isn't any particular anatomy lessons I learned that day, but instead something presumably

R. Rosenberger, *Against Frog Dissection*,
https://doi.org/10.1007/978-3-032-14346-4_3

tangential to the activity itself: I remember the packaging. The frogs for dissection for my class all came packaged together in a large, vacuum sealed, clear plastic bag. I don't recall whether the sealed bag that I saw was for the activity I was about to take part in, or for another class after mine. But I won't ever forget the sight of all those dead frogs stacked across one another. It was impossible to tell exactly where one frog ended and the next started, overlapped haphazardly as they were, held in place by the stiff and transparent plastic.

Let's step back for a moment and think about classroom dissection in literal terms. That is, rather than think about the dissection activity in all the ways that it is commonly understood (e.g., as a learning activity, as a science-like experiment, etc.), let's take a moment to just describe the thing that students encounter there on the desk or table in the classroom. To put it plainly: students are presented with a dead frog.

In most cases, the frog that is presented to students is specifically *already dead*. (Of course, there are stories in which live frogs are brought to class to be killed there by the teacher or even the students, like the ones freed by Elliot and his alien friend in *E.T.*). That is, the frog first encountered by many students is a cadaver, killed already, elsewhere and at some other time.

Continuing with this literal description, we can note that the frog is also *already in the classroom*. That is, the frog presented to students has been removed from its environment, transported from there, and brought to the student, ready-to-go. There is often a preservation process involved. Formaldehyde or some other substance may be utilized to keep the corpse from degrading. Norm Friesen summarizes the process as follows:

> it is important to note that the animal in this dissection has itself been deracinated or uprooted from its original context. However, for this radical de-contextualizing to have been successful, a laborious and protracted effort is required: the animal's body has to be drained of its natural fluids, and these have to be replaced with preservatives; the body needs to be kept in a sealed container, and once the dissection has begun, further steps need to be taken for its preservation. (2011, p. 101)

In this way, the frog that is presented to the student is one that has been isolated from its ecological context, treated, transported, and is first encountered by students in the context of the classroom.

And finally, we can note that the frog for classroom dissection is not encountered alone. It is presented *already in the context of the activity*, introduced along with the other tools of dissection, such as the wax tray and the pins that students will push through the arms and legs to hold the frog body in place, as well as the forceps and knife. The frog is encountered by students not only in the context of the classroom, but also in the context of the dissection activity. The frog is encountered *as a frog-for-dissection.*

I take these observations of the literal situation of classroom dissection—that the frog is already dead, already preserved and packaged and transported, already in the classroom—to be obvious and uncontroversial. However, I suggest that when we highlight these literal aspects of the frog dissection situation, then we are prompted to think about it differently. We can follow this out and, rather than conceive of this activity in terms of all the things associated with the dissection paradigm discussed in the previous chapter, we can also think about the frog dissection situation as a concrete material practice. That is, frog dissection is not merely a tradition, a biology lesson, or an encounter with nature, and all that. The frog that is presented to students is also a specific material object, and one that has been altered and used by people.

My claim is that it can be useful to think about the dead frog of the classroom dissection activity *as a technology*. That is, we should approach our study of classroom frog dissection here in terms of how the dead frog of this classroom dissection activity is itself an artifact, a material "thing" that humans have made in a particular manner, an object that humans are using in particular ways for particular purposes. This opens up specific questions about how the dead frog can and cannot be used, what specific educational opportunities it affords and disaffords, and how exactly it does in fact tend to be used in practice. As we'll see in this chapter, philosophers of technology and others claim that our technologies influence and shape our experiences in many ways. If we conceive of the dead frog as a technology, then this opens up questions about exactly how this technology influences and shapes classroom experience. And my suggestion is that approaching the dead frog itself as a technology is one of the best ways to upend the dissection paradigm.

There are of course a number of reasons why it might at first seem counterintuitive to approach the frog-for-dissection itself as a kind of technology. For one thing, it might seem like a category mistake. Isn't a frog a natural object, not a human-made artifact? Isn't the whole point of the frog dissection activity to explore the natural world, with the frog as the sample of the natural world? As the philosopher of technology Don Ihde observes, one popular belief is "that technology by being produced is *artificial* and the artificial is to be contrasted with the *natural*" (1990, p. 6). Following Ihde and others, we can challenge this intuition. That is, we can challenge the idea that "the natural" and "the technological" are separate and mutually exclusive categories. As Ihde continues, "Stone, once shaped, enters praxis. Hardness becomes hardness for _____; shapability, shapability for _____. Thus, what is 'natural' in the stone becomes artifactual within the relativity" (1990, p. 70). In this way, in the case of the frog dissection activity the inertness and dissectability of the dead frog body becomes a frog-for-dissection.[1]

The frog presented to students in the classroom is thus different from one that someone might stumble upon when, say, taking a stroll by a pond. You may see a frog hop into the pond as you walk by. Although it would be unusual, you could even come across the corpse of one lying in the path. As noted above, this is distinctly different from the experience had by the student who encounters the frog-for-dissection, an animal that has been snatched out of its natural habitat and transported to the classroom, killed for this purpose, and presented in the context of the dissection activity, ready-to-go. Sure, the frog-for-dissection remains in many ways a natural object, the corpse of something that had hatched from an egg and matured through natural processes. But it is additionally something that has been altered by human hands in ways that are necessary for its availability as part of a classroom activity, and in ways that are consequential to the learning experience. These consequences, as we'll see, are both good and bad.

Another reason that it may be counterintuitive to conceive of the dead frog as a technology is because it may at first be strange to think of the frog itself as a kind of tool. Isn't the dead frog the object of investigation?

[1] Ihde and others join a tradition of thinkers in the fields of philosophy of technology and Science and Technology Studies that challenge the common conception of the world as something whose parts are separable into a series of exclusive binary categories, such as nature vs. culture, or the scientific and technological vs. the social and political, or even objective vs. subjective (see, e.g., Latour 1993; Haraway 1997; Verbeek 2011). According to these thinkers, the overreliance on these kinds of binaries leads to many confusions and is a contributor to many contemporary problems. Here, I am adding the ethical, environmental, and—as is the specific focus of this book—educational problems of frog dissection to this discussion.

Isn't it a scientific specimen? Isn't the whole point that we're using science to explore the world through the dissection activity? It is important to think of classroom frog dissection not only as a science-like investigation, but also as a concrete practice. And in this sense the dead frog must itself be understood for what it is within this activity: a learning instrument, an apparatus taken up as a means for teaching particular biology lessons, a classroom technology. The frog-for-dissection is similar in many ways to other pedagogical tools, such as chalkboards, desks, computers, Bunsen burners, globes, calculators, art supplies, or microscopes.

One more reason that it may seem counterintuitive to think of the dead frog as a kind of technology is that the frog is conventionally contrasted with alternatives such as computer simulations of dissection. Isn't the computer simulation the key technology at issue here? Isn't the frog the real thing, and the computer simulation the artifact? Of course, the computer simulations and other alternatives to frog dissection are technologies. And in the case of simulations in particular, there are consequences—both advantageous and disadvantageous—to the fact that they are things entirely created by human hands from the ground up. However, the status of the dissection alternatives as technologies does not change the dead frog's status as a technology as well. It is important to conceive of *both* the frog-for-dissection and its simulated alternatives as technologies.

My claim here is that reassessing the dead frog in terms of its status as a technology is important for reconsidering the dissection paradigm. As discussed in the previous chapter, the practice of educational frog dissection is one that is ensconced snuggly within a number of associated conceptions: the frog is real, and the simulations are not; frog dissection is akin to a science experiment; frog dissection is the educational ideal, and simulations should only attempt to imitate it; and so on. If we want to critically analyze frog dissection, then we need to find ways to think about it separately from these gathered associations. I suggest that when we highlight frog dissection as a concrete material practice, we can analyze it in its specificity. Frog dissection can be contrasted with its alternatives in terms of their advantages and disadvantages as educational tools.

This makes space for considering the practice of frog dissection in terms of the analytical perspectives of the humanities and social sciences. In particular, in this chapter, we'll reconsider frog dissection in terms of insights from the field of philosophy of technology and related lines of thought.

And more, I suggest that the dead frog itself should be analyzed as the *guiding technology* at issue. That is, when we consider the practices of those parts of the world where the dissection paradigm holds, the frog-for-dissection itself is the technology whose presence chiefly influences the situation. The centrality of the frog dissection activity sets the context for everything else related to it.

In what follows in this chapter, I apply some ideas from the philosophy of technology and related fields to the topic of frog dissection. First, we can consider frog dissection as a form of what some scholars refer to as "technological mediation." That is, we can consider how the technology of frog dissection changes and informs the classroom situation, how it implicitly assigns the various people involved (the teacher, the students) into particular roles, and how it sets up the activity as a whole as a particular kind of learning experience. Next, we'll consider the social situation of frog dissection. With ideas from the social studies of technology, we can explore how disagreements over the design and purpose of these technologies of biology education have settled into a particular form in the case of the dissection paradigm, and how the reasoning surrounding this debate has become calcified in specific terms. And, lastly, we can explore the concrete human-technology relationships of the frog dissection activity. This all sets us up to appreciate the contingency of the dissection paradigm, and it opens up ways to think about things otherwise.

3.1 The Technological Mediation of the Dead Frog

When you put a technology into a situation, it changes things. Think about the act of driving a car. The car makes new things possible. A person with a car can of course travel greater distances at faster speeds. But it is not just our abilities that change. Driving comes with various tradeoffs. You need to learn to drive. You should in most cases stay on the roadways and obey the various signals and signs and lights. The car needs gas, or another power source. There are environmental tradeoffs. And so on. But there's still more: the car somehow reshapes our situation. It assigns various roles and duties and meanings. A person becomes a driver. They become responsible for the driving situation, one with rules and stakes. Another person becomes a passenger. Someone walking outside becomes a pedestrian, or maybe even a jaywalker. The whole situation falls under the authority of traffic law. And so on. A question for us here is: how does the technology of the dead frog reshape the educational situation?

A challenge in the study of technologies is to find ways to think about how they not only make new opportunities possible, but how they also influence our actions and decision-making along the way. The technologies we use of course do not force us to take them up in a particular manner. Or, put in the language of the philosophy of technology, our devices themselves do not "determine" exactly how they should be designed, used, and understood. However, in addition to rejecting this kind of technological determinism, we should also reject the idea that technologies are conversely somehow entirely neutral. Although they do not fully determine our circumstances, they also do not completely abstain from influencing our situation; technologies are "non-neutral" contributors. A central claim I am attempting to articulate here in this book is that by putting aside the idea that the dead frog is merely an encounter with the real world, and by conceiving of it explicitly as a technology, we are prompted to think about these non-neutral contributions.

I suggest that one way to begin thinking about the dead frog's status as a technology is to conceive of it as a form of "mediation." Peter-Paul Verbeek, a leading thinker on these issues, observes that, "Technologies help to shape the way in which reality is present to human beings; not only how they perceive the world, but also the frameworks in which they interpret it" (2005, p. 122). Following thinkers like Don Ihde, Verbeek and others explore how technologies can be understood to mediate our relationships with the world around us.[2] According to this line of thinking, as technology usage changes a person's relationship with the world, that person and the world itself are also changed along the way. As Cathrine

[2] These ideas build from a philosophical perspective called "postphenomenology." This perspective provides a kind of practical tool kit for helping to articulate and analyze people's experiences with technology (e.g., Ihde 2009; Rosenberger and Verbeek 2015). A central line of this research explores the "co-constitution" or "coshaping" of "technological mediation," the kind taken up in this section (e.g., Verbeek 2005; Verbeek 2011; de Boer 2020; Hasse 2020; Kudina 2023; Rosenberger 2024). Postphenomenology's commitment to understanding the co-constitution of subjects and objects puts it in a kinship with other theoretical perspectives, especially feminist accounts of materialism and posthumanism (e.g., Haraway 1997; Barad 2007; Braidotti 2013). We can note as well that other scholars make use of the terminology of "mediation," but in different, though related, ways (e.g., Latour 1999).

Hasse puts it, "technologies are neither good nor bad per se, but implicated in human embodied relations to the world… In the mediated relationship, subject and object become something new in the relation" (2020, p. 253). In this way, we can understand technology usage to be a kind of mediation that, under this terminology, "coshapes" or "co-constitutes" users and the world. As Verbeek writes, "Mediation does not simply take place between a subject and an object, but rather coshapes subjectivity and objectivity… Humans and the world they experience are the products of technological mediation, and not just the poles between which the mediation plays itself out" (2005, p. 130). When we think about technologies in terms of the notion of mediation, it prompts us to explore how the various related people and objects and institutions of the world are coshaped in the process.

For example, in perhaps the leading case study in this line of research, Verbeek considers the ways that the technology of obstetric ultrasound co-constitutes the context of childbirth (2011). The use of sonography to provide real-time imagery of a developing human fetus can be considered a mediating technology. Of course, this technology has importantly changed our ability to monitor the health of a fetus within a mother's uterus. The mediation of this technology can also be understood to radically transform the context of pregnancy in multiple ways, with, as Verbeek argues, moral repercussions. The view afforded within the sonogram casts the fetus as a kind of individual, an at least somewhat autonomous figure onscreen. The mother is correspondingly transformed into a kind of environment for the fetus, one that can be more or less hospitable. The mediation of the technology thus co-constitutes everyone's various roles. The pregnancy itself is more deeply established as a medical situation, one that falls under the expertise and authority of doctors. And within this medical situation, the fetus is afforded the role of a kind of patient. In addition, according to Verbeek, the technological mediation of the ultrasound co-constitutes the moral situation of pregnancy. This technology provides access to information about the fetus's health that informs a potentially fraught decision-making process, one whose options include pregnancy termination. The parents themselves are transformed into moral decision makers.

Verbeek additionally observes that the choice to forgo doing an ultrasound is one that is also itself co-constituted by this technological mediation. That is, even the decision *not* to use ultrasound during a pregnancy is today one that takes place in a specific moral context. The choice to not make use of an ultrasound falls within the decision-making framework set by this central guiding technological mediation; choosing to opt out is itself a moral decision.

Or take, for example, Bas de Boer and Olya Kudina's exploration of the technological mediation of machine learning algorithms in the practice of medicine (2021). Machine learning is used especially in medical imaging practices, and is taken up for making predictions and diagnoses regarding everything from irregular heartbeats to breast cancer development. As de Boer and Kudina note, this kind of technology usage "is celebrated for reducing the supposedly subjective aspects of diagnosis by medical practitioners and for increasing the trustworthiness of the diagnoses given insofar as it enables medical judgments that purportedly rely on the objective outcome of algorithmic processes" (2021, p. 246). However, these technologies have also given rise to several fears, such as the concern that machine learning tools could become overly relied upon, or that their luster of mechanical objectivity could lead people to forget just how deeply the outputs of such algorithms are informed by what specific data are fed into them. De Boer and Kudina reframe this situation by considering how the technological mediation of machine learning has the potential to transform medical diagnostics. They write, "Using ML [machine learning] in medical practice affects the observation capacities of medical professionals in the process of diagnosis. Through the presence of ML, medical professionals, patients, and the relationships between them are co-constituted in new ways" (de Boer and Kudina 2021, p. 250). For example, these devices change what it means to be a medical practitioner, a job that has increasingly become centered on interpreting algorithmic outputs (and, potentially, navigating the lack of transparency of some algorithmic processes). These devices also change the nature of medical responsibility, as part of the authority for making diagnoses is handed over to machines. And there are specific new concerns that arise, such as the potential for medical diagnoses to become based primarily on what kinds of information it is possible to enter into the algorithm, rather than on other potentially relevant forms of data.

We can look to everyday technologies as well. Think again of the example of driving. In my own work, I have been investigating the dangers of using the phone while behind the wheel (Rosenberger 2024). It turns out that decades of scientific research in the field of cognitive science have shown that using the phone while driving is dangerously distracting. This has been found to be the case not only for texting (which is by far the most dangerous option), but also for holding a phone conversation. And, perhaps most interestingly, this has been found to be equally true for both handheld and hands-free calling. So, consider the design of many contemporary cars that include high-tech dashboard infotainment systems with touchscreen and voice-interactive interfaces, as well as advanced "connectivity" features that pair with a driver's smartphone to enable many options for hands-free calling, texting, and other forms of communication. It is true that the availability of these features does not somehow force drivers to use them. They don't fully "determine" the situation in a way that compels a person to drive in a dangerously distracted manner. However, I suggest that the design of contemporary dashboards anticipates a particular kind of driver, one that is not fully paying attention to the road. Contemporary dashboards incline distracted driving.

This point was underlined for me in my experience purchasing my first new car. All the vehicles I had owned before had been "used" cars. At the car dealership, all the stressful haggling and paperwork had been completed, and our new car was brought out to us. As a final task, the salesmen asked for my phone. They wanted to "pair" it with my dashboard. But I declined. They were baffled. However, for me in that moment, in this particular material and social arrangement, in the midst of this form of technological mediation, the decision to forgo using my phone while driving was not merely a choice not to do something. It was something that needed to be actively refused.

I argue that the dead frog should be understood as the central mediating technology of the dissection paradigm. The activity of dissecting a dead frog—as a concrete material practice—serves as the guiding technological mediation of this situation. The frog corpse as a specific "thing" along with its accessories in the dissection activity such as the forceps and scalpel together co-constitute the biology lesson itself, its participants, and even the dissection alternatives.

The technological mediation of the dead frog puts into place the roles played by the various human participants. The teacher becomes a kind of lab director, an overseer of the classroom which itself is co-constituted into a laboratory-like context. The students themselves are put into the role of quasi-scientists, or perhaps, scientists in training. In addition, those students who make the choice to opt out of the frog dissection activity are put into the role of conscientious objector. That is, even those who choose not to take part in the activity are co-constituted within this form of technological mediation. The choice to opt out is established as one based on personal moral objection. And this choice is couched not as someone simply selecting one option rather than another; it is a form of refusal.

We can note that the alternatives to frog dissection are themselves also quite deeply materially co-constituted by the technological mediation of the dead frog. Dissection alternatives tend to take the form of an activity that mimics the action of frog dissection. As noted, in the case of computer-simulated alternatives, we often see a virtual frog onscreen, as well as a virtual scalpel controlled by a mouse or touchscreen interface. Physical simulations often involve an artificial frog body that students slice apart with a knife. Arguments made in favor of such simulated alternatives hold that these reproductions of the central elements of the frog dissection activity are good enough that they can pass as adequate stand-ins. In this way, the dissection alternatives are deeply co-constituted by the technological mediation of the dead frog, including everything from the concrete details of their design to the justifications made for their adoption.

The technological mediation of the dead frog may also come to co-constitute the biology lesson itself. The strength of the dissection paradigm may influence not only how these biology lessons are taught, but also what they comprise. That is, considering how deeply some parts of the world appear to be in the grasp of the dissection paradigm—as evidenced, for example, by its status as a cultural tradition or a rite of passage, and by the mimicry its alternatives are forced to implement—the mediation of the dead frog may come to influence the pedagogy. The lesson itself becomes, at least in part, whatever it is that is taught by dissecting a frog.

The dividing line between what is real and what is artificial is something that itself can become co-constituted through technological mediation. Frog dissection is a prime example of this. The mediation of the dead frog establishes itself as the real thing at issue, at least within the practices and discussions surrounding this activity. Although it is in many specific ways an artifact prepared by humans, the dead frog is presented as a specimen of nature, an object of the actual world selected for examination. And the frog-for-dissection is cast against the simulated alternatives. Since the simulated alternatives are built from the ground up, they are unequivocally artificial things, cultural artifacts, human-made devices, and they produce an experience of something "virtual." The frog-for-dissection becomes defined in this comparison as the real thing, the natural thing, something into which we can peer and make discoveries about the world.

And, lastly, the science-adjacency of the dissection activity itself is something that is co-constituted by the mediation of the dead frog. Through the technological mediation of the dead frog, this classroom learning activity becomes associated with the practice of science. That is, this activity becomes something that is not only used to teach a science lesson in the classroom; in perhaps a partial or restricted way, this activity postures as an opportunity for students to *do* science. As the students themselves are co-constituted as quasi-scientists, and as the frog itself is co-constituted as something "real" and "natural," the activity takes on the connotations of scientific research, things like "objectivity" and an engagement with the "facts" of the world. The act of doing science, as it is taught to young people through this activity, is co-constituted as one in which things like any potential personal discomfort or any sentiment for animal life should be compartmentalized, overcome, or set aside. Through the co-constitution of the frog dissection activity as something science-like, the implied lesson is made that science involves desensitization and the instrumentalization of animal life.

I suggest that these connotations that get attached to the dead frog are part of what can make it so counterintuitive to reconceive of it as a technology, something made for particular human uses, in accord with particular human values, and with particular consequences, both moral and pedagogical.

Many features of the dissection paradigm may often in practice simply be taken as givens. It is often taken as given that the real and the virtual are two separate and pre-existing categories, and that the frog-for-dissection goes in the former and simulated alternatives in the latter. It is often taken as given that the biology lessons for which frog dissection is optimally suited are the best and only ones that should be taught through this kind of activity. However, when we recognize that the dead frog can be approached as a form of technological mediation, we open the door to challenging these assumptions. When we recognize that the dead frog is a technology that mediates the classroom situation, we can draw out many ways that this thing itself influences our conceptions of the real and virtual, and how it shapes what its associated lessons have become. And this opens up questions about how things could be otherwise. As Verbeek suggests, "The central *teleological* question in technological culture, to conclude, is 'what kind of mediated subjects do we want to be?'" (2011, p. 84).

3.2 The Social Construction of the Dead Frog

An important aspect of technology to keep in mind is that any device is the result of a history of work done by other people. The anthropologist of technology Bruno Latour urges us to, "Think of technology as *congealed* labor" (1999, p. 189). When we encounter a technology and deal with it, we are doing so in some relationship with the work of other people performed at some other time and some other place. That work is, in some form, enacted on us as the person currently dealing with the technology, whether we're interacting with it with our bodies (like when we drive a car) or we are perceiving and interpreting it (like when we see the traffic light at the intersection). This is one of the major things that perspectives on technology from the field of Science and Technology Studies (STS) help us to do: reveal and analyze the human labor that has made our current technological situation possible. You and your car may together drive down the roadway to get to your destination. But that action is made possible by lots of work done in the past by others. That work includes any number of things, from the design and manufacture of that car, to the programming of the traffic lights, to the social negotiations that have led to car culture and the roadway system.

Another advantage of approaching the dead frog as a technology is that it sets us up to think about it in terms of its social situation. Rather than conceive of it as an ideal, or as something merely real to be discovered, we can consider the frog dissection activity as something that is the result of social negotiations. Ideas from the field of STS are useful for drawing out this kind of social labor. This can help to denaturalize an object of study. That is, these perspectives can help to undercut intuitions that the things of the world are merely natural objects, or are devices with inevitable essential characteristics, or that their design has been determined exclusively by a kind of technical logic. These perspectives can be useful in situations in which the thing under examination is typically conceived as if it could only ever be one way. In order to reveal the contingency that underlies the dissection paradigm, we can explore how it has been socially constructed.

One useful term coming out this literature is what Trevor Pinch and Wiebe Bijker have referred to as "interpretive flexibility." They write that,

"By this we mean, not only that there is flexibility in how people think of, or interpret, artefacts, but also that there is flexibility in how artefacts are designed. There is not just one possible way, or one best way, of designing an artefact" (Pinch and Bijker 1984, p. 421). This notion is helpful for thinking about the way that the main design and purpose of a technology is something that can be more or less socially settled. When a technology is first introduced, there may not be general agreement about how it should be used or how it should fit into society, and thus, also, how it should be further developed. Different social groups may emerge with differing conceptions of the device, and different plans for it. In such a case, we could say that there is a high degree of interpretive flexibility. The notion of interpretive flexibility is useful for challenging intuitions and arguments that hold that technological development reduces to a kind of technical logic. As Bijker puts it, "The concept of interpretative flexibility is crucial in countering technical determinism. Indeed, to recognize the interpretative flexibility of artifacts is synonymous with refuting technical determinism" (1993, p. 118). Of course there are technical dimensions to how technological development occurs. But technological development is also led by social forces that guide how that development should proceed.

The interpretive flexibility associated with a technology decreases as the matter becomes more and more socially settled. In this terminology, social debate over the meaning and usage of a technology may come to a close over time. As Bijker explains it, "Closure, in the analysis of technology, means that the interpretative flexibility of an artifact diminishes. Consensus among the different relevant social groups about the dominant meaning of an artifact emerges" (1995, p. 86). Over time, a community can even come to forget there was ever disagreement over a technology in the first place. With this reduction in interpretive flexibility, the device may come to be understood as normal, as just the way things are. However, it is also of course possible for this social situation to change. As Bijker continues, "It is important to recognize that the process of closure is almost irreversible—almost, but not completely" (1995, p. 87).[3]

[3] These ideas build from a perspective called the "Social Construction of Technology" (SCOT) (e.g., Pinch and Bijker 1984; Bijker et al. 1987; Bijker 1995; Pinch and Trocco 2002; Oudshoorn and Pinch 2003). As a central account of technology coming out the field of STS, SCOT offers a framework for conducting empirical investigations into the social factors that influence technological development. This kind of sociological, anthropological, and historical work proceeds, generally speaking, through the identification of

Think of early bicycles, such as the Penny-farthing, the kind with the enormous front wheel popular in the mid-to-late-1800s. Before learning about the work of Pinch and Bijker on this topic, I had assumed that one could draw a line from these old-timey bikes straight through to contemporary bicycles. Or more precisely, I hadn't thought about it much at all, so I held an unreflexive assumption that if we looked at the main definitions, designs, and purposes of today's bikes (e.g., they have a front and back wheel of a similar size), then we'd find the big-wheeled historical bikes to simply be a less well-developed technology in these same terms. That is, I thought the old, big-wheeled bikes were merely less advanced versions of the same thing. It turns out to be more complicated than that.

Pinch and Bijker instead show that the bicycle has a branching history, one that features radical disagreements about their meanings and design (e.g., Pinch and Bijker 1984; Bijker 1995). The historical bikes with a large wheel had an element of adventure and danger to them, which makes sense when you think about it; the rider sat very high off the ground. With these kinds of bicycles, cycling took on what Pinch and Bijker refer to as a "macho" connotation, an activity for men, especially athletes. However, there were also others, such as women and older men, who thought that the high-wheeled bikes were unsafe and impractical for usage as a mode of transportation. In this way, the bicycle at that time was subject to substantial disagreement, with differing plans for how the next versions of cycling should be developed. As Pinch and Bijker write, "Thus the interpretative flexibility of the artefact 'Penny-farthing' is materialized in quite different design lines" (1984, p. 424). From here different styles of bicycle develop, and the problems that the different groups see with bicycle development change. As bicycle wheels come down to the same size, and as tires are introduced, the bicycle evolves into a design more akin to what we have today. The athletes grow to focus on the speed and racing enabled by these developments, and those with safety concerns grow more satisfied with these more physically stable models.

Or think of the technology of oral contraception. As it exists today, the birth control pill is something that women take to prevent pregnancy. Before learning about the work of Nelly Oudshoorn on this topic, I had assumed that there were straightforward biological reasons why the pill is available for women and not for men. Or again, more precisely, I hadn't

the social groups relevant to disagreements within a community about a technology, and the investigation of the issues involved in the settlement of those disagreements.

thought much about the history of this device, and had always unreflexively assumed that the lack of a pill for men was simply a matter of biology. I had assumed that, since women have babies, we had figured out how to use hormones to interrupt that process in a way that was not as straightforward a possibility for men. Oudshoorn shows that this is simply not the case. As she puts it, "the constraints on the development of new contraceptives for men cannot be ascribed to intrinsic properties of bodies or technologies" (Oudshoorn 2003, p. 225).

The idea for an oral contraceptive pill for men has long been in development, and could be a real option today if not for a history of resistance coming from multiple directions, including a historical lack of investment by the pharmaceutical industry, among other factors (Oudshoorn 2003). This "asymmetry" in practices of contraception, as Oudshoorn refers to it, has persisted in large part due to an entrenched cultural conception of masculinity, one that sees it as inconsistent with any responsibility for family planning. As Oudshoorn writes, "The predominance of modern contraceptive drugs for women has disciplined men and women to delegate the responsibility of contraception largely to women. In the latter half of the twentieth century contraceptive use became excluded from hegemonic masculinity" (2003, p. 232). Oudshoorn's work follows the worldwide efforts to develop male hormonal contraceptive technologies, and the headwinds they continue to face. An emerging theme is that any progress in this area has required work to shift conceptions of masculinity along the way. She writes, "In the development of new contraceptive technologies for men, the construction of masculinities is at the forefront of the design. Technological innovation thus becomes a process of designing technology and masculinity" (Oudshoorn 2003, p. 16). As these long overdue developments continue to advance, they will need to be accompanied by continuing changes in our conceptions of men as clinical trial participants, as family planners, and as sharers of the risks and responsibilities of contraceptive technologies.

My suggestion is that it is useful to reconceive of the frog-for-dissection as a technology, and thus as something open to the kinds of social analyses offered here. This could help to denaturalize frog dissection. That is, these ideas help to challenge the intuition that the predominance of the practice of frog dissections in school in those places where it is common is exclusively due to any kind of technical determination, or to the intrinsic properties of bodies. Reconceiving the frog-for-dissection in these terms helps to highlight how it could be otherwise.

In particular, I recommend that we think of the dissection paradigm as a kind of cultural disagreement which has come to a close in a particular way. The technology of the dead frog, I suggest, is these days one with a low degree of interpretive flexibility. This is evidenced by the strength of the hold of the paradigm, the massive numbers of frogs killed annually for this activity, its place as a cultural touchstone, its pervasive representation in media, etc. This is evidenced as well by the design of alternatives, which are relegated mainly to imitating this activity, rather than do something else.

Of course, disagreement remains. As noted, one way that issues around frog dissection are often cast is exactly as a debate. However, I contend that the social settlement of the dissection paradigm includes the establishment of a debate that exists with particular terms. In general, this debate has become set within a context that generally assumes several specific background positions: (1) the frog dissection activity is the standard; (2) alternatives must set out to reproduce the dissection experience; and (3) it is sometimes acknowledged that individual moral objections, while exceptions, should be accommodated. This, for example, is the opinion conveyed by the position statements released of the American biology teacher organizations mentioned in the previous chapter. This is the implicit understanding conveyed by student choice laws that require that concerned students have the opportunity to opt out of the dissection activity and learn these same lessons in another manner. In this way, I suggest that the contemporary social consensus is upon a debate with this specific structure. Of course, we can find critics of frog dissection who attempt to argue differently. However, this specifically structured "debate" can be conceived as the thing that today, within the dissection paradigm, maintains a strikingly low degree of interpretive flexibility.

I suggest that the explicit recognition of this structure is the first step in the work of dismantling it. As Bijker notes, it is a difficult task to reintroduce interpretive flexibility after community disagreements have closed. As Oudshoorn shows, this kind of work includes not only arguing for one side of a disagreement, but also for both changing how we understand certain basic concepts, perhaps how some people think of themselves and their roles, and even how things are designed. Nevertheless, this is all exactly the kind of work I would like to put on the table. In addition to waging the debate over frog dissection as it has been waged—with critics refining their ethical, environmental, and educational arguments—I hope that more explicit work can be done to call attention to and undermine the established consensus of the dissection paradigm itself.

3.3 The Zero-Sum Logic of the Dissection Paradigm

Another way that we can approach the socially settled nature of the dissection paradigm is in terms of what we could call a kind of shared pattern of reasoning. That is, we can think about a "common sense" that structures discussions around the frog dissection activity. Is there a general backgrounded and assumed reasoning behind the dissection paradigm?

My suggestion is that the dissection paradigm can be characterized as structured by a specific reasoning pattern, one in which it is assumed that any digression from the basic activity of dissecting a dead frog in the expected manner must involve an educational sacrifice. This is a part of what it means, in the context of the dissection paradigm, for the frog dissection activity to have taken a place as an assumed ideal. I suggest that the socially settled logic of the dissection paradigm is such that any alternatives to the ideal of the frog dissection activity are understood to necessarily result in a loss. The philosopher of technology Andrew Feenberg has criticized this kind of thinking, writing that, "design is not a zero-sum economic game but an ambivalent cultural process that serves a multiplicity of values and social groups without necessarily sacrificing efficiency" (1999, p. 95). Following Feenberg here, let's refer to this pattern of reasoning at work in the dissection paradigm as a kind of zero-sum logic. It is a socially settled pattern where it is assumed that, in the rivalry over what is understood here as good education, there can only be one winner. Any alternatives, by this logic, must then be losers to some degree. That is, under this reasoning, some educational value must be lost when an alternative is taken up. Perhaps there may be a valid *external* reason to do so, such as accommodation for a student's individual moral preferences, as the dissection paradigm sometimes allows. But such an accommodation according to this zero-sum logic, will involve at least some sacrifice to the quality of education. Hopefully, as this reasoning goes, the alternative provided for this accommodation will at least be educationally adequate.[4]

We can push back against this kind of "commonsense." As Feenberg continues, "Raising the standards means altering the definition of the object, not paying a price for an alternative good or value" (1999, p. 95). Sure, there are good and bad designs, at least in the way that some are better suited to some purposes than others. But we should be suspicious when a pattern of reasoning takes hold within a community that claims that it is impossible to innovate solutions to problems without a sacrifice. Sometimes sacrifices are required to solve a problem, of course. But opponents to potential innovations may attempt to undermine proactive solutions by calling upon a zero-sum logic that is not a necessary feature of all design. We should be especially suspicious when opponents appeal to zero-sum reasoning in response to concerns raised by others about ethics, equality, or justice.

To better articulate the ways that a faulty zero-sum logic has been used to defend the status quo against those who argue for design alterations for the greater good, let's consider a few examples regarding advocacy and the economy.

One of Feenberg's go-to examples comes from the history of child labor in the United States. It was common for children to work in places like factories as recently as the 1930s. Feenberg notes that back in those days, "Proposals to abolish child labor met with consternation on the part of factory owners, who regarded the little worker as an 'imperative' of the technologies created to employ him" (2010, p. 12). That is, child labor was understood at that time in terms of a zero-sum logic. Sure, its defenders reasoned, there may be concerns about child welfare to be considered, but child labor cannot be eliminated without destroying the economy. These defenders could point to tangible things in the world: factory technologies were designed with the smaller bodies of children in mind. All those machines would need to be remade. Even if the removal

[4] In particular here, I draw on Feenberg's notion of the "tradeoff model" (1999, p. 94; 2010, p. 33). He takes up a similar and overlapping set of ideas to the ones reviewed so far in this chapter, and uses them to explore the ways that societal hierarchies become entrenched in design and culture. The reason I am not using the "tradeoff model" language in this chapter is because I lean on the word "tradeoff" centrally in the next chapter, and there with a slightly different connotation. So, here I'll use the zero-sum terminology to emphasize the dissection paradigm's assumption that the accommodation of moral considerations necessarily always involves a sacrifice of educational ones.

of children from the workforce serves some important value, they argued, such a change would necessarily come with an unacceptable economic cost.

However, as we see in US history, the outlaw of child labor did not result in economic disaster because the economic system changed in other positive ways. With children out of the workforce, they could instead be educated, and they could go to work after childhood with better skills and preparation. As Feenberg puts it, "Children ceased to be workers and were redefined socially as learners and consumers. Consequently, they entered the labor market with higher levels of skill and discipline that were soon presupposed by technological design" (2010, p. 12). These social transformations to the economy have changed how we think of everything from the nature of labor to what it means to be a child. Feenberg writes,

> In developed countries, child labor violates fundamental assumptions about the nature of childhood. Today we see children as consumers, not as producers. Their function is to learn, insofar as they have any function at all, and not to earn a living. This change in the definition of childhood is the essential advance brought about by the regulation of labor. (2010, p. 39)

Against a zero-sum reasoning that assumed a ban on child labor would inevitably result in economic loss, we see that in this case innovations within society made it possible to both retain economic functioning and at the same time live up to our moral obligations to children.

Another example can be found in discussions on environmental sustainability. One common criticism of advocacy efforts is the claim that any help to the environment comes at the direct expense of the economy. Conservation, as this criticism of environmental advocacy goes, means we will have less. Such criticisms thus take up a zero-sum logic; one thing necessarily comes at the expense of the other. Feenberg disputes this reasoning. He writes, "Environmentalism will not impoverish our society. We will go on enriching ourselves, but our definition of prosperity and the technologies instrumental to it will change and become more rational in the future judgement of our descendants" (Feenberg 2010, p. 43). Within work on environmental advocacy, I've always thought that the notion of "sustainability" was especially strong for disrupting this zero-sum economic logic.

A sustainable practice is one that can continue being performed indefinitely into the future (e.g., McDonough and Braungart 2013; Norton 2005). Through innovation and study, we are developing resources that can be used over and over into the future, as well as jobs that can be done generation after generation. This could be achieved in many ways, such as when the resources utilized are renewable, or when processes regenerate what they use, or when practices have been calibrated such that they utilize a resource up to and only up to a point that it naturally replenishes itself (like in the case of sustainable fishing). Advocates for sustainability thus tout *both* its environmental *and* economic benefits. For example, Mindy Lubber, the President and CEO of the sustainability advocacy non-profit organization, Ceres, put the point this way in a testimony to a subcommittee of the US House of Representatives in 2024, "Climate change is an issue that presents both serious financial risks and opportunities for investments and job creation in the rapidly emerging clean energy economy. Investors and companies are acting on climate because the business case is compelling."[5] With this style of innovation, we may be able to counter a zero-sum logic that holds that we always must decide between either the environment or the economy.

I suggest that the zero-sum logic of the dissection paradigm presents us with a false dilemma: We must choose between valuing the education of children or valuing animal life. Under this logic, any alternative to the activity of dissecting a dead frog necessarily comes at the expense of educational value. By identifying and challenging the dissection paradigm, we are challenging the assumption that alternatives to frog dissection must always by some necessity result in an educational cost. It is possible instead that they may offer something equally good, or perhaps even better, but different.

3.4 User Relationships with the Dead Frog

Speaking as someone living within one of those countries where the dissection paradigm prevails, I'd describe the state of affairs like this: there is an air as if this all has always been inevitable. The issues here are settled and configured in a way that they can seem obvious and necessary. The default consensus seems to be that since frog dissection

[5] See: https://assets.ceres.org/files/testimony-of-mindy-lubber,-ceres-inc.-20240612.pdf.

provides a memorable hands-on learning experience, it should be the standard way its associated lessons should be taught, perhaps with the occasional proviso that students with personal moral objections can opt out. With its posture as an educational ideal, and as an encounter with something natural and real (rather than merely artificial and virtual), and as a science-like endeavor, frog dissection can seem unassailable, at least from a vantage point inside the paradigm. I want to challenge all of this.

In this chapter, I have argued that one way to clear a path for reanalyzing frog dissection is to explicitly conceive of the dead frog *as a technology*. We should approach frog dissection as a concrete material practice. However else we think of it, we should also recognize the frog-for-dissection for what it is: a thing, an object manipulated and changed by human hands, killed and shipped, bought and sold, and used as a classroom pedagogical tool. This prompts us to analyze it through the lenses of philosophy and social theory, as we would any other technology. We can reveal its role as a form of technological mediation, and consider the ways in which it co-constitutes its own situation within the dissection paradigm as real, and natural, and science-like, and as an educational ideal. We can explore the socially constructed status of the dissection paradigm, and consider what it might take to reintroduce a level of interpretive flexibility into our understanding of the technologies of these biology lessons. Rather than approach the technology of the dead frog as necessary, innocent, and inevitable, we can see this learning situation as something that could be otherwise.

If these thoughts clear a path for thinking about frog dissection differently, then what would it mean to proceed down that path? One way forward is to think about frog dissection and its alternatives in terms of their usage. As concrete material things, we can think about them in terms of the specific ways they make possible a transformed relationship to our world. As Don Ihde claims, "*for every revealing transformation there is a simultaneous concealing transformation of the world, which is given through technological mediation*. Technologies transform experience, however subtly, and that is one root of their *non-neutrality*" (1990, p. 49). We can consider the non-neutral ways in which the technologies of the dead frog and its alternatives make possible particular relationships with the world generally, and with the classroom lessons of biology in particular.

I suggest that a way forward is to follow Ihde and think about technology usage in terms of tradeoffs. As a concrete material practice, the dissection of a frog corpse offers specific learning opportunities. In addition, it may retain distinctive limitations, or introduce specific problems. Simulations too, of course, retain specific limitations. They also may afford specific learning opportunities different from those offered by the hands-on dissection of a dead frog. And more, by unshackling simulated alternatives from the assumption that they must mimic frog dissection and must be evaluated primarily in terms of how well they do so, we can begin brainstorming ways that dissection simulations could be developed and expanded to better capitalize on their own distinctive educational opportunities.

We'll explore these various limitations and opportunities in the next chapter.

References

Barad, Karen. 2007. *Meeting the Universe Halfway: Quantum Physics and the Entanglement of Matter and Meaning*. Durham: Duke.

Bijker, Wiebe E., Thomas P. Hughes, and Trevor J. Pinch, eds. 1987. *The Social Construction of Technological Systems: New Directions in the Sociology and History of Technology*. Cambridge: MIT Press.

Bijker, Wiebe. 1993. "Do Not Despair: There Is Life After Constructivism." *Science, Technology, & Human Values* 18 (1): 113–138.

Bijker, Wiebe E. 1995. *Of Bicycles, Bakelites, and Bulbs: Toward a Theory of Sociotechnical Change*. Cambridge: MIT Press.

Braidotti, Rosi. 2013. *The Posthuman*. Cambridge: Polity.

de Boer, Bas. 2020. *How Scientific Instruments Speak: Postphenomenology and Technological Mediations in Neuroscientific Practice*. Lanham: Lexington Books.

de Boer, Bas, and Olya Kudina. 2021. "What Is Morally at Stake When Using Algorithms to Make Medical Diagnoses? Expanding the Discussion Beyond Risks and Harms." *Theoretical Medicine and Bioethics* 42 (5–6): 245–266.

Feenberg, Andrew. 1999. *Questioning Technology*. New York: Routledge.

Feenberg, Andrew. 2010. *Between Reason and Experience: Essays in Technology and Modernity*. Cambridge, MA: MIT Press.

Feenberg, Andrew. 2017. *Technosystem: The Social Life of Reason*. Cambridge, MA: Harvard University Press.

Friesen, Norm. 2011. *The Place of the Classroom and the Space of the Screen: Relational Pedagogy and Internet Technology*. New York: Peter Lang.

Haraway, Donna J. 1997. *Modest_Witness@Second_Milennium*. New York: Routledge.

Hasse, Cathrine. 2020. *Posthumanist Learning: What Robots and Cyborgs Teach Us About Being Ultra-Social*. New York: Routledge.

Ihde, Don. 1990. *Technology and the Lifeworld*. Bloomington: Indiana University Press.

Ihde, Don. 2009. *Postphenomenology and Technoscience: The Peking University Lectures*. Albany: SUNY Press.

Kudina, Olya. 2023. *Moral Hermeneutics and Technology: Making Moral Sense Through Human-Technology-World Relations*. Lanham: Lexington Books.

Latour, Bruno. 1993. *We Have Never Been Modern*, trans. Catherine Porter. Cambridge: Harvard University Press.

Latour, Bruno. 1999. *Pandora's Hope: Essays on the Reality of Science Studies*. Cambridge: Harvard University Press.

McDonough, William, and Michael Braungart. 2013. *The Upcycle: Beyond Sustainability—Designing for Abundance*. New York: North Point Press.

Norton, Bryan G. 2005. *Sustainability: A Philosophy of Adaptive Ecosystem Management*. Chicago: University of Chicago Press.

Oudshoorn, Nelly, and Trevor Pinch, eds. 2003. *How Users Matter: The Co-Construction of Users and Technologies*. Cambridge: The MIT Press.

Oudshoorn, Nelly. 2003. *The Male Pill: A Biography of a Technology in the Making*. Durham: Duke University Press.

Pinch, Trevor J., and Wiebe E. Bijker. 1984. "The Social Construction of Facts and Artifacts: Or How the Sociology of Science and the Sociology of Technology Might Benefit Each Other." *Social Studies of Science* 14 (3): 399–441.

Pinch, Trevor, and Frank Trocco. 2002. *Analog Days: The Invention and Impact of the Moog Synthesizer*. Cambridge: Harvard University Press.

Rosenberger, Robert. 2024. *Distracted: A Philosophy of Cars and Phones*. Minneapolis: Minnesota University Press.

Rosenberger, Robert, and Peter-Paul. Verbeek, eds. 2015. *Postphenomenological Investigations: Essays on Human-Technology Relations*. Lanham: Lexington Books.

Verbeek, Peter-Paul. 2005. *What Things Do*. University Park: Penn State University Press.

Verbeek, Peter-Paul. 2011. *Moralizing Technology*. Chicago: Chicago University Press.

CHAPTER 4

The Tradeoffs of Frogs and Simulations

Abstract If we follow the previous chapters and set aside many of the common presumptions about classroom animal dissection, then this practice can be evaluated instead in terms of its concrete educational advantages and disadvantages. In particular, the often overlooked educational limitations of the dissection activity can be explored, and specific educational problems can be highlighted. In addition, if we can move beyond presumptions that dissection simulations must mainly aim to mimic the animal dissection activity, then novel lines of educational development can be identified. These include the potential for dissection simulations to expand into lessons on animal habitat, cell structure, and evolutionary history, or to integrate scientific imaging readouts and other media, among other possible expansions.

Keywords Frog dissection · Educational technologies · Biology education · Simulation · Animal welfare · Humane education · Technological mediation · Philosophy of technology · Postphenomenology

What happens when we pluck the activity of frog dissection out of its various cultural associations, hold it up, and examine it as a learning experience? What happens when we release simulations of dissection from the obligation to primarily attempt to reproduce the experience of cutting

R. Rosenberger, *Against Frog Dissection*,
https://doi.org/10.1007/978-3-032-14346-4_4

apart a dead frog? Put in the language of this book, what happens to *both* the frog dissection activity ***and*** dissection simulation when they are removed from the dissection paradigm? When recast in these terms, how does this change how frog dissection and dissection simulations stack up against one another?

As I've argued in the previous chapters, under the dissection paradigm the tendency is to consider dissection simulations as an "alternative" to be judged against frog dissection as an assumed ideal. The act of cutting up a dead frog is held as the standard, and simulations are evaluated in terms of how well they measure up to that standard. This evaluation is then conducted primarily in terms of how closely or how poorly the simulation reproduces any of the experience of slicing open a frog cadaver.

When extracted from the context of the dissection paradigm, we can instead contrast frog dissection and dissection simulations simply as two educational activities. We can evaluate both of them individually as different concrete material practices. We can explore both as specific classroom activities that offer different educational possibilities.

Put differently, we can make this contrast in terms of tradeoffs. That is, we can approach both frog dissection and dissection simulations as classroom tools that make it possible to educate in specific ways and that simultaneously cut off particular educational possibilities. We can explore them in terms of the specific educational benefits and limitations distinctive to each. And when evaluating both frog dissection and dissection simulations, we can contrast the specific advantages and disadvantages of each against one another.

To be clear, the task of contrasting the tradeoffs of these activities is a non-neutral affair. In light of the prevalence of the dissection paradigm, the project of comparing the distinctive educational advantages and disadvantages of frog dissection and dissection simulation involves the upending of received wisdom. That is, the task does not merely include enumerating and juxtaposing the particular things that each of these activities can and cannot do; it involves exploring how some things may be overvalued, as well as how others may be overlooked. Since the dissection paradigm consists of associations and biases in favor of the frog dissection activity and against simulated alternatives, the work of this comparison involves actively undercutting some of these associations and biases.

In this way, in what follows a special emphasis will be placed on exposing some of the distinctive shortcomings of the frog dissection activity since these can go overlooked when it is conventionally conceived as an ideal. And more, we'll need to highlight specific ways that the frog dissection activity is educationally problematic, again because these problems can become obscured when this activity is itself assumed to be the standard of comparison.

In addition, in what follows it will be important to put special emphasis on the potential for simulations to be expanded beyond their conventional formats. That is, the distinctive advantages of simulations tend to get overlooked in these discussions insofar as they tend to be evaluated primarily in terms of how well or poorly they meet the standards set by the frog dissection activity. And more, we will need to brainstorm new ways that such simulations could potentially be expanded to better capitalize on these overlooked advantages. The use of simulations for teaching anatomical lessons is an unmet potential.

4.1 The Limits and Problems of Using Dead Frogs as Educational Tools

The educational advantages of frog dissection are well known. That is, with the dissection paradigm exerting a comfortable hold upon those places where it prevails, the educational advantages of this activity are an explicit part of the discussion. Even still, these distinctive educational benefits associated with the act of cutting apart and exploring a dead frog are worth going through again here. As noted, the overarching advantage of this activity is its hands-on and visceral quality. Frog dissection involves physical engagement with a tangible object: pinning, and handling, cutting and snipping, pulling apart and dismantling. And these physical engagements are multiply stimulating. The slices and scissor cuts are met with resistance from the cadaver. For many students, the view into the cadaver will be surprising, constituting new sights and sounds. It is smelly. The vivid, and visceral, and potentially queasy aspects of this experience make it a distinctive one compared to the others that students tend to have in the classroom on a typical day.

All of these distinctive characteristics of the frog dissection experience—its vividness, and grossness, and tactility, etc.—are inextricably tied to the other centrally touted educational advantage: it is memorable. Since the experience of frog dissection is so viscerally impactful, the activity is often characterized as unforgettable. Defenders point to these qualities as educational benefits. If the experience is vivid and memorable, then the associated school lessons should also get etched into a student's memory along the way. Defenders can appeal to the ideals of progressive education, such as those advocated for by John Dewey and others. The frog dissection activity creates an engaging experience in the classroom, one that links to issues of practical concern, including the basic functioning of bodies. So, all of this may help to cement lessons and connect those teachings to students' lives. (As we'll see, simulation defenders can similarly make an appeal to progressive education ideals).

Defenders can point to additional advantages as well. For example, it is sometimes argued that the one-time quality of the dissection matters; once a cut is made, it cannot be undone. There may be pedagogical value to the care required to make the incisions correctly the first time, and to the irreversible consequences to any mistakes. In addition, defenders could argue that there is a pedagogical value to the dissection of frog bodies as things developed through natural biological processes. That is, there could be value in the encounter with a genuine and flawed example of a creature's body. This can be contrasted with idealized models of frog bodies, such as a diagram in a textbook, or a simulated frog designed by people.

With these kinds of advantages noted, we can also explore the educational limitations and problems associated with the frog dissection activity.

4.1.1 *The Limitations of Dead Frog Bodies*

Let's next consider some of the limitations of the frog dissection activity. As a concrete material practice, this activity is of course saddled with specific educational drawbacks. As a tool with particular characteristics, the dead frog can be used for some educational purposes and not others. It opens the door to some educational possibilities, and closes others. These closures can be highlighted as limitations or tradeoffs, and, when held in comparison to alternatives, cast as drawbacks or disadvantages.

We can begin by thinking about the inertness of the dead frog, its lack of dynamism. The frog dissection activity presents students with one moment of the frog's life. As a part of making a frog-for-dissection, it becomes fixed, as it were, in that moment. The frog body is preserved to prevent decay. This situation, by necessity, presents only a snapshot. Students are presented with the frog body in the moment of its life when it was killed. If we think of this limitation as a disadvantage, then this sets us up to consider potential alternatives to this activity, ones that could enable the exploration of an animal's anatomy in more than a single moment of its life. (In the next section, we'll consider alternative activities that could potentially integrate lessons about the various stages of a frog's lifespan).

We can also think about this inertness in terms of the bodily motility of the dead frog. Of course, the frog corpse is changeable and manipulable in specific ways: it can be posed, and bent, and sliced, and dismembered, and otherwise taken apart and explored as part of the dissection activity. However, there are limitations to this motility that must also be noted since they restrict which kinds of educational affordances are possible. For example, as an inert mass, this object cannot offer an exploration of the bodily movements of a living frog. A living frog's bodily motions are of course interesting and distinctive and relevant to frog physiology, such as how they hop, how they swim, how they stretch their tongue, how they reproduce, how they hibernate, etc. All of these topics are cut off by the specifics of the frog dissection activity. That is, anything related to the kinds of motions of the frog's life are inaccessible here since someone has already killed it.

The internal systems of the dead frog are equally inert. Since the frog-for-dissection has been killed, it does not breath. Its blood doesn't flow. Its digestion has stopped. These things are obvious, of course. But they have educational consequences. Again, these limitations can prompt us to consider how alternatives might be able to include, rather than inherently omit, educational experiences related to the dynamics of internal bodily systems. (And again, we'll do just this in the next section in our exploration of potential areas of expansion for dissection simulations).

An additional characteristic of the frog-for-dissection is that it is presented to students specifically in isolation from its natural surroundings. As has been noted, rather than encounter the living frog in connection with its environment, the frog dissection activity specifically presents an already dead specimen prepared as an experiment-like learning activity. This arrangement of course presents specific educational opportunities. However, we can note that it at the same time also smuggles in particular limitations. The presentation of the frog-for-dissection in terms of a laboratory-like context is one that cuts off the possibility of exploring the frog in terms of its connections to its natural habitat, its ecosystem. This, again, prompts us to consider alternatives that could instead attempt to do just that, even in a simulated manner, as we'll do next in the next section.

This is all to say that the concrete material circumstances of frog dissection commit this activity to being one that only explores lessons of anatomy through an inert and isolated mass, rather than as a dynamic and living being. Or, put another way, it is a *choice* of pedagogical design to explore the interior of the frog body through means of dissection. It is a choice to utilize a frog-for-dissection as an educational tool, something already killed, already separated from its environment, already set up like a scientific experiment. This is a choice made in the face of other options for exploring issues of anatomy, such as imaging, or simulation. This choice has consequences for what can be taught and learned and what cannot. Of course, lessons on these topics that are structurally omitted by the frog dissection activity can simply be taught separately. However, it is important not to overlook the fact that the frog dissection activity does not lend itself to teaching these lessons on anatomy *in connection* with issues of the frog's bodily motions, the dynamics of its internal systems, and its relationships to its habitat.

4.1.2 The Problems of Dead Frog Bodies

In addition to being limited in particular ways, the frog dissection activity may also be specifically problematic. When thinking about the tradeoffs of frog dissection, it is important to consider specific problems introduced by this activity. Of course critics of frog dissection argue that this practice is morally problematic. However, here I want to focus on the educational problems of frog dissection. Does the frog dissection activity also implicitly teach bad lessons, lessons that are morally wrong?

The principles of progressive education offer insight here. As noted in the first chapter, the frog dissection activity can be seen in some ways as a paradigmatic example of a form of progressive education in the Deweyan sense: this activity creates a vivid experience in the classroom, one that could potentially generate connections to students' lives. However, these same goals of creating classroom experiences should also direct us to be mindful of the details and the implications of the experiences we create. As Dewey writes, "Perhaps the greatest of all pedagogical fallacies is the notion that a person learns only the particular thing he is studying at the time" (1938, 48). The experience created by the frog dissection activity may be vivid and memorable and hands-on. It may be genuinely instructional. However, it may additionally impart lessons on more than the stated goals of biology instructors. And it is possible that some of these lessons taught by implication are bad ones, things we shouldn't be teaching.

Here, I'd like to explore some problematic lessons imparted by frog dissection that fall into two categories: (1) the value of animal life, and (2) the role of the scientist. That is, we'll first consider morally problematic lessons that instruct a disregard for animal welfare. And, second, we'll explore a problematic portrayal of scientific practice as something that essentially includes desensitization. Finally, we'll consider how the materiality argument helps to put these problems into relief.

First, problematic lessons on the value of animal life. I have argued that in those moments when the issue of frog dissection is cast as a debate, the dissection paradigm tends to divide the arguments for and against frog dissection in a particular way: defenders of the *educational value* of frog dissection go on the one side, and critics of the *ethics* of frog dissection go on the other. Let's revisit these ethical criticisms here. I would like to highlight the way that one of these moral criticisms cut across this divide. We can identify *moral problems with the education provided by the frog dissection activity*.

Generally put, the central moral argument against frog dissection is that it involves a disregard for animal life. While this could be approached in different terms (e.g., a violation of animal rights, a weighing of ethical costs), critics argue that it is unethical to kill these animals merely for their use in a classroom activity, especially when we remember that the associated lessons can be taught adequately, if not equally well or perhaps even better, through alternative means. These concerns are multiplied when we recall that massive numbers of animals are killed for dissection across the world annually. These concerns are magnified further when we consider the ethical consequences regarding not only the treatment of those animals themselves, but also the environmental impact that results from the mass removal of these animals from the ecosystem. It is not my goal here to evaluate these moral arguments (even if, personally, I am sympathetic to them, and if, ultimately, the materiality argument developed in this book can be used to bolster these ethical criticisms). However, there is an *educational concern* inextricably related to these moral concerns. For anyone who does not entirely discount these points about the value of animal life, for anyone who believes the value of an animal's life to be above zero, it must be asked what is taught by this activity—at least implicitly—about the proper treatment of animals. Does the frog dissection activity provide a kind of implicit moral instruction? If so, then are the morals that are taught problematic ones?

That is, the particular argument at issue here is that frog dissection is bad because it teaches students problematic lessons about how animals should be treated. For example, F. Barbara Orlans notes that it could be argued that "rather than teaching respect for life, dissection cheapens it. It can be very confusing to students to be taught on the one hand to care for animals and revere life when on the other hand they see destruction of life in their biology classes" (1988, 39). Or as Jan Oakley puts it,

> Dissection's hidden curriculum teaches lessons about the ethics of animal use: specifically that there is little value in animal life. Although its stated intent is to teach students about biology, dissection also teaches that it is ethically acceptable to kill animals in the name of school. (2013, 7)

According to these criticisms, the frog dissection activity teaches that it is okay to hurt and kill animals, at least if it is done for schooling or science, and even if these scientific and educational justifications are thin. It teaches, at least implicitly, a kind of animal cruelty.

Second, problematic lessons on the role of the scientist. A second educational concern to be raised about the frog dissection activity is that it may impart problematic lessons about science itself. Insofar as frog dissection is cast as a science experiment and as a chance to actually do laboratory science, this specific learning experience may imply problematic things about what work in science involves. Insofar as the frog dissection activity comes across to students as representative of science as usual, then students are given a depiction of what it means to be a scientist. This prompts an educational question: What account of science itself does the frog dissection activity provide, even implicitly?

Of course, this activity portrays the work of science as including many positive and uncontroversial things, such as care and precision, rigorous note taking, detailed observation, etc. However, as noted, a central part of the experience of many students participating in animal dissection activities is a feeling of discomfort, queasiness, and disgust, which can include a moral revulsion. Some defenders even point to this discomfort and disgust as exactly part of what makes the experience so memorable. Part of the process of doing the frog dissection activity can thus also include becoming desensitized, or learning to compartmentalize these feelings, or otherwise getting over them. Jane Goodall herself puts it this way:

> By and large, students are given the implicit message that it is ethically acceptable to perpetuate, in the name of science, a variety of unpleasant procedures against animals. They are encouraged to suppress any empathy they may feel for their subjects, and persuaded that animal pain and feelings are of a different nature from our own, and that there is little value in animal life. (2002, vii)

Some teachers may even consider this desensitization to be an expected part of the experience. Students may receive the message by implication that science necessarily includes an instrumental attitude toward animals. And it can be argued that it is problematic if we are teaching our children that an important part of being a scientist includes things like desensitization, compartmentalizing moral qualms, and the instrumentalization of the lives of animals.

Critics fear that this implicit instruction may have the effect of turning some students off to science. Students may decide that if this is what science is like, then it is not for them. Dorian Solot, whose anthropological work on exactly these issues was mentioned in chapter two, notes that, "One of the skills learned from dissection is to see the animal as a scientific tool rather than a once-living being... For some girls, especially, this transition is difficult and unpleasant. The message is that detachment from caring about animals is a key skill for scientists. If you don't have it, science may not be the right field for you."[1] In this way, the frog dissection activity could serve a kind of gatekeeping function, problematically discouraging some students from developing an interest in science.

These educational problems of the frog dissection activity are emphasized in particular ways here in terms of the materiality argument. While the preceding chapters have not been specifically about ethics (we're not doing deep ethical theory in these pages), we need to reconsider these particular ethical issues because they are inextricable from the educational tradeoffs introduced by the concrete material situation of this activity. My contention is that the materiality argument advanced throughout this book is useful for drawing out the details of the problematic lessons imparted by the frog dissection activity. Both parts of this argument—(1)

[1] Quoted in Averett, Nancy. (2020). "High School Dissections Are a Science Class Tradition. But Are They Doing More Harm Than Good?" *Discover*. Jan 20th, 2020. https://www.discovermagazine.com/the-sciences/the-argument-against-high-school-animal-dissections.

the identification of the dissection paradigm, and (2) the reconceptualization of the dead frog as a mediating technology—help to highlight aspects of the implicit instruction of problematic lessons by the frog dissection activity.

First, the dissection paradigm. The above reflections on the patterns of the dissection paradigm help to explain how these problematic lessons can become downplayed in conventional discussions on the topic. One factor is the tendency for the frog dissection activity to be cast as the standard. Insofar as frog dissection is approached as itself the definition what should be taught and how this pedagogy should occur, this of course disallows any suggestion that whatever is taught could somehow be a problem. Another aspect of the dissection paradigm that can function to downplay these problematic lessons is its tendency to cast the ethical issues of frog dissection in terms of students' personal morality. Insofar as issues of ethics in these discussions are relegated to an individual's stance regarding the killing of frogs and/or the effects on the environment, these issues can appear to refer only to things outside of the learning experience, such as issues of animal sourcing. Here instead we see that ethical issues run through the education of frog dissection itself.

An additional aspect of the dissection paradigm that may contribute to the downplaying of problematic lessons imparted by frog dissection is this activity's association with scientific experimentation. Insofar as this activity tends to be portrayed within the dissection paradigm as an opportunity for students to actually do science by performing a quasi-experiment in the classroom, frog dissection can become associated with a naïve notion of scientific objectivity, one that holds that anything scientific is inherently separate from people's subjective perspectives, motivations, social agendas, or moral concerns. An air of scientism around the frog dissection activity may help to conceal the possibility that it could even implicitly teach something morally wrong, or something wrong about how to think of the work of scientists.

Second, the dead frog *as* a mediating technology. Another way to highlight elements of these problematic lessons is to approach frog dissection as we did in chapter three: as a concrete material thing. I suggest that we can draw out these problematic implications by thinking through the frog dissection activity not as an abstract idea or plan, but as it is actually experienced in practice in the classroom.

We can think, for example, about students' experiences at the beginning of the activity. I had mentioned in chapter three that a noteworthy part of my own experience of high school frog dissection included a memorable and unsettling encounter with the pile of dead frogs, packaged and vacuum sealed. We can follow this out to consider the ways that presentation matters. The lessons of frog dissection may begin before the first slice. How are the cadavers revealed, handled, and provided to students?

These questions extend through participation in the activity itself. It is easy to imagine cases of students goofing around and doing things to or with the corpse that in other contexts could be considered disrespectful. Does this matter? In certain bald consequentialist terms, perhaps it does not. If the particular frog-for-dissection is already dead by the time it gets to the student, then the animal itself is not put in pain by any classroom misbehavior. Nor does this classroom behavior impact the number of frogs killed for this activity. However, we can consider the harm in terms of personal virtue, that is, in terms of how actions train one's individual character. What makes such bad behavior "bad" isn't merely that the student may not be following classroom rules or the instructions of the activity, nor is it merely whatever bad effects are caused. These behaviors are bad for the small way that they train one to be a worse person, insofar as they do.

Here's the important point. Thinking about the moral training experienced by the misbehaving student should prompt us to consider something more core to the frog dissection activity: the character training experienced by the student who performs the activity properly and straightforwardly. There is a kind of character training that happens through this performance too. Insofar as there are problematic lessons taught by implication about the significance of animal life, and insofar as there are problematic lessons taught about how scientists instrumentalize animal life, these things can be understood not only as information communicated implicitly by the activity, but also as values trained through participation. This activity, this learning-through-doing, includes a process of desensitization. The act of participating in the frog dissection activity constitutes an instance of practice, of practicing actions in tune with particular values, of character building. These problematic lessons are baked in as part of the value training that occurs through the memorably impactful activity of animal dissection.

We can think as well about the conclusion of the activity, and the moments after. What happens in the classroom when students have completed the dissection activity? And what values are communicated by the material practices of these final moments? We can keep an eye on what happens when the student is finished. Perhaps they toss the cadaver in the garbage. Disposing of the cadaver, however that is done, may be part of how students clean up their workstations. Or perhaps the cleanup duties fall to the teacher, and the student gives it no further thought. The trashcan is thus a non-innocent material player in the implicit training of these problematic lessons.

It is helpful to approach frog dissection as something that involves tradeoffs because it highlights the educational pathways that this activity opens up, and also those it closes off. Given the strength of the dissection paradigm, the educational pathways opened up by the frog dissection activity are well known; they have come to define what it is that we should want this kind of activity to accomplish. This is why the tradeoff language is useful. It highlights that not only does the frog dissection activity make some educational benefits possible, like hands-on exploration. It also closes off, or disaffords, some possibilities for pedagogy. It is no coincidence that the dissection paradigm downplays the things that the frog dissection activity closes off. As we'll see in the next section, these are the same educational opportunities that simulation can open up again.

4.2 Expanding Dissection Simulation

The educational disadvantages of simulations are well known. That is, with the dissection paradigm exerting a comfortable hold on those places where it prevails, the educational disadvantages of dissection simulations are an explicit part of the discussion. Even still, it is worth reviewing these dynamics here for the sake of setting up an exploration of how simulations of frog dissection could be expanded to better achieve their educational potential.

As noted throughout the book, the central argument against taking up simulated alternatives to frog dissection is that they cannot replicate the experience of cutting up an actual dead frog. That is, even though dissection simulations may ultimately be used to teach the same biology lessons at least equally well, they cannot reproduce that same experience of physically slicing open an animal cadaver, with all its goo, and physical resistance, and abjection, and stink. Under the thinking of the dissection paradigm, this failure is pivotal; since the dissection paradigm sets the standard for the success of a simulation at whether or not it can reproduce this exact experience, computer simulation's inability to do so clinches the debate.

Here's the thing: as mentioned at the outset of this book, I find part of this to be convincing. I agree that computer simulations cannot merely reproduce the exact experience of cutting a cadaver. Computer simulations with their static images, videos, and 3-D models onscreen, as well as their keystrokes and touchscreens and mouse clicks, cannot deliver the same experience as an in-person and hands-on encounter with a dead frog. Even physical simulations, which of course come a lot closer, cannot reproduce this experience exactly in all its slimy and stinky discomfort. I disagree, however, with any suggestion that this situation should close the debate over which of the two—frog dissection or simulated dissection—is the preferable educational option. It is only if we blindly adopt the perspective of the dissection paradigm that the dead frog appears to be the obvious preference. As I've tried to show throughout this book, the setup of the dissection paradigm is contingent and socially situated. Things could be otherwise. *We should not accept the stipulation, no matter how widely assumed it may be, that nothing can equal the educational benefit of dissecting a dead frog.* We should instead remain open to the possibility that simulated alternatives provide at least the same educational benefits. And we should remain open to the possibility that simulated alternatives could be developed further.

This brings us to a central piece of the argument of this book. If we can distance ourselves from the dissection paradigm, then we can look at dissection simulations differently. That is, if we can release dissection simulation from the charge to mainly mimic the activity of cutting up a dead frog, then we could consider new ways to design simulation-based educational activities on these same topics. And we could better highlight the ways that contemporary dissection simulations often already include educational features that go beyond the mere mimicry of frog dissection.

Approached as concrete material practices, dead frogs and simulated alternatives are not the same. They maintain different strengths and weaknesses, different affordances, different potential connections to other things. Dead frogs afford hands-on manipulation. They afford a quasi-experimental setting of a particular type. Simulation instead affords flexibility and customizability. The experience of a computer simulation is directed to the thing onscreen, rather than to the hands-on and faux-laboratory context. Computer simulation may not lend itself to reproducing a hands-on laboratory-like experience, but it is also not limited to this aim. It need not posture as an experiment-like activity at all. Importantly, what is available onscreen is radically open to modification, so the connections that simulation makes possible are wider than and not limited to those afforded by the dead frog. Simulation need not even pose as an activity where the user approaches something dead with the task of cutting it open. With these framings set aside, simulations could be expanded to include connections to a variety of other relevant areas of biology, as we'll explore below.

There is one more thing to note before brainstorming the ways that dissection simulation could be expanded: it is important to resist getting too flowery with our descriptions of simulated experience. In sticking with a phenomenology of the usage of these learning tools, my goal in what follows is to describe experiences that expanded simulations could make possible. It is tempting to use linguistic flourishes for such cases, for example, to muse about how a computer simulation could include views at a microscopic level and enable students to zoom through a capillary, or something like that. It's true that simulations could include views into microscopic perspectives on the body, and that is a promising possibility that we'll think through further below. However, if we want to stay true in our descriptions of a student's actual experience, it is also important to remember that these viewpoints are merely made available onscreen, engaged through mouse clicks and touchscreen interface. So, at the risk of weighing the following descriptions down, I'll try to stick to a language of "virtual frogs" and other things simulated "onscreen."

Let's consider a few of the various dimensions upon which simulated dissection could be expanded.

4.2.1 Spatiality

We can think about the spatial situation of the typical simulated dissection program and how that may be expanded. It is understandable that the virtual context that is displayed onscreen within dissection simulations typically takes the form of a laboratory-like experiment, or even a classroom laboratory, often with a simulated wax tray, pins, etc. Given the demands of the dissection paradigm, such a virtual setup is prudent. That is, since—within the dissection paradigm—simulated dissections must justify themselves as adequate alternatives by mimicking the frog dissection activity as closely as possible, it makes sense that a simulation would offer a lab or classroom context onscreen as part of this implicit justificatory argument.

However, when we set aside these artificial demands of the dissection paradigm, simulation can be expanded into different contexts. In particular, there is the opportunity to connect these anatomical lessons to relevant education about the animal's environment. The student's explorations of the frog's body and its internal systems could be performed *in connection* with an exploration of the specifics of the frog's habitat, the pond and mud, the food chain, the changing seasons, or even how changes in that environment have bodily effects on the virtual frog. Or, to put it in more literal terms, these elements of biology education can be taught through an exploration of the virtual frog's relationship onscreen with its simulated environmental niche. Rather than teach a more limited set of lessons on anatomy in isolation from the animal's natural surroundings, as in the case of the faux-lab setting of the frog dissection activity, these same lessons could potentially be taught more wholistically. Lessons on the animal, its body, and its internal workings could be taught in terms of their part in the ecosystem.[2]

4.2.2 *Motility, Interiority, and Magnification*

Computer simulation makes it possible to look at a virtual frog specimen in multiple ways that are impossible for the dead frog. As a human-created program, we can design from the ground up exactly how we'd like students to peer inside the frog onscreen. Some of these potential perspectives offer distinctive educational opportunities.

To begin, a virtual frog could be examined in motion. As noted, the dead frog cannot jump, breathe, or swim as would a living frog. The virtual frog is also not a living being. However, unlike the dead frog, the virtual frog onscreen need not lie there inertly. The frog onscreen could potentially move about, and in numerous ways. These motions could be potentially interactive, responding to student inputs as they engage with the computer interface. As noted, a frog's bodily motions are interesting and distinctive. The frog's hop, its sticky and out-flipping tongue, or its croak or chirp could all be simulated digitally. (So, for example, rather

[2] For example, the Digital Frog 2.5 dissection simulation, defunct these days but once the gold standard, included lots of information about frog ecology, as well as information on different frog species.

than only encountering the vocal cords inside a dead frog's throat, a simulation could demonstrate them in motion and even provide samples of the sounds they generate). In combination with the points made above about simulated spatiality, the motility of the anatomy of the virtual frog could be explored in terms of its relations to its virtual environment, its external and jelly covered frogspawn, the camouflaged skin of some or the colorful poison announcement of others, their defensive secretions, their swimming, and singing, and tree climbing, and so on.

Computer simulations enable unconventional perspectives into the interiority of the virtual frog's body. Of course, physical dissection activities do this: the frog-for-dissection is to be sliced open; the physical simulated frog is similarly dissected with a knife; physical anatomical models of a frog are to be opened and disassembled. However, computer-simulated frogs could potentially be inspected layer by layer. Different internal systems—nerves, skeleton, organs, musculature, etc.—can potentially be viewed in isolation from one another or in conjunction, rendered as 3-D models onscreen.

Computer simulations have even been at times criticized for being too seamless and unrestrictive in the way a bit of the virtual frog body can merely vanish when clicked. For example, Douglas Allchin writes that, "A proper aim is discovery, not destruction. One should separate and clarify: Trace pathways, find boundaries, encounter connections—quite impossible if things are pre-cut and disappear as preformed units in a single mouse-click" (2005, 270). While this feature of computer-simulated frog dissection can be cast as a disadvantage when contrasted with the challenge and resistance imposed by the dead frog body, this same feature can also be appreciated for its potential educational benefits.

For example, a 3-D model of a particular bodily system could potentially be displayed onscreen, rotated, and observed from angles otherwise difficult or impossible to explore within the body of a frog-for-dissection. The viewpoint onscreen could potentially include zooming capabilities, enabling students to magnify particular areas. And combining these points with the above ones on motility, we can note that these simulations of isolated interior systems could potentially be observed *in motion*. For example, a simulated circulatory system could potentially include the movement of digital blood through its vessels. That is, the virtual frog could potentially afford students and teachers the opportunity to explore not just what an organ looks like, or where it is positioned within the body. *Simulation could enable the exploration of internal bodily systems in their functioning*: blood flow, digestion, respiration, etc. Several contemporary dissection simulations offer just these kinds of views of isolated systems in motion.[3]

We can expand further on the issue of magnification. In this brainstorm of directions available for expanding simulation, another possibility is the action of zooming further and further inward. A classroom anatomy computer simulation could potentially include exploration of bodily processes that occur at the microscopic level. So, for example, a computer-simulated frog dissection could potentially include examination of the physiology of cellular processes. The point is that, *unlike the dead frog, simulations are not inherently limited to the scale of the full frog body, or even the scale of the limits of human vision.* A simulation could potentially include investigation into virtual cells, say, muscles cells, or blood cells. Once again, the simulation could enable a view onscreen into the dynamic processes happening at this level. These investigations could potentially connect together, enabling students to, say, follow air into the frog's lungs, zoom in to see the process of oxygen intake into the alveoli, and zoom in further to see its transport through the blood and delivery to other cells. On this topic of frog respiration in particular, there are once again specific interesting things to note, such as the fact that frogs additionally engage in cutaneous respiration, taking in air through their skin. This sets up potentially interesting comparisons for students to make with human breathing. The various combined dynamical, magnified, and distinctive perspectives that simulation could make possible introduce novel pedagogical possibilities, ones with great potential for encouraging students to synthesize information on multiple important biological processes.

We can also think about how these kinds of expansions could apply to physical simulations of animal dissection. It makes sense that physical simulations of frog dissection often take the shape of a frog body lying ready to be cut open; such a strategy directly neutralizes what it often touted as the main distinctive advantage of the activity of dissecting a

[3] Some examples of contemporary dissection computer simulations that make it possible for students to view isolated systems within the frog in a 3D manner include Biosphera's 3D Frog Anatomy, Visible Body's 3D Frog Model, and Embibe's Froggipedia.

See: https://biosphera3d.com/product/3d-frog-anatomy-software/.

See: https://www.visiblebody.com/blog/no-frog-no-prob-a-3d-visual-guide-to-frog-anatomy#:~:text=When%20the%20frog%20is%20in,Ribbit%20ribbit.

See: https://www.embibe.com/exams/froggipedia/.

dead frog, its hands-on quality. However, if we go beyond the dissection paradigm, then we can explore additional possibilities for teaching anatomy through physical dissection simulation.

For example, it is possible to imagine physical dissection simulations of various sizes. Physical simulations could be enlarged to provide students with a zoomed-in view of simulated organs. Instead of (or in addition to) dissecting a frog-sized simulation and encountering a life-sized simulated frog heart, students could also be presented with a large physical simulation of a heart, with explorable and potentially dissectible ventricles, etc. Physical simulations could include versions that isolate particular systems, enabling students to, say, explore and handle a model of the frog's circulatory system or skeleton. And while it may involve considerable complexity or expense, we can even imagine the possibility of physical dissection simulations that include some forms of motion, such as bodily motion, or the motion of interior systems, such as the circulatory system. In these kinds of ways, physical simulations too could be developed in ways that go beyond attempts to merely reproduce the frog dissection experience complete with that activity's own educational limitations.

4.2.3 Temporality

Another advantage of computer-simulated dissection is its potential to display a virtual frog that changes over a span of time. Simulations could be expanded further in terms of this temporal dimension. As noted, the dead frog has a distinct and static temporal situation. Where the dead frog is limited to enabling exploration of only one moment of the sample's life—the moment it was killed—simulation is not saddled with this same restriction. Computer simulations could potentially be expanded to include additional moments, including information about a frog's seasonal rhythms, life cycle, and even species history.

For example, a simulation could be expanded to include interaction onscreen with the virtual frog at different parts of its yearly life. If the particular frog species simulated onscreen happens to be one that undergoes hibernation in wintertime, then this could be included in the learning experience. The virtual frog could be explored onscreen in its changing relationship to its virtual environment, for example, with the inclusion of information about the frog's bodily torpor when it has buried itself under the mud or at the bottom of the pond. Within a simulation, the virtual frog's position in time across the year could be potentially toggleable. A simulation could include the ability to compare internal processes as they occur at different times, presented onscreen in motion (e.g., blood flow), enabling a view of both hibernative and active states. (And it is possible to imagine the kinds of activities that could accompany this simulation usage, such as ones that encourage students to think about changes in that occur in their own bodies when asleep). These anatomical lessons—impossible, we can note, in the activity of dissecting the dead frog—could potentially be included in a simulation, and, depending on its sophistication, in an integrated, dynamic, and interactive fashion.

The introduction of a temporal dimension into simulation makes it possible to consider including lessons on the frog's life cycle. Of course, frogs undergo an interesting set of metamorphoses across their lives, hatching from eggs, and going through stages of maturation as they grow from a swimming tadpole to an adult frog, sprouting legs and losing their tail and gills. The flexibility of computer simulations enables the possibility of integrating anatomical lessons about maturation and metamorphosis into the activity. Such lessons could include things like 3D animations, slideshows, or timelapse video. Contemporary simulations are already exploring these possibilities.[4]

The introduction of a temporal dimension into simulation even makes it possible to consider the evolution of frog species. Again, frogs are an interesting topic in this regard. They emerged on Earth around 200 million years ago, during the age of the dinosaurs. The extinction of the dinosaurs itself may have contributed to the explosion of frog species we see today.[5] Simulation could include information about the emergence of frog species from earlier amphibians during the Mesozoic Era (the same era in which mammals also first appear), and the later proliferation of species across the globe. Put differently, it is possible to imagine simulation expanded to include not only lessons on contemporary frog anatomy, but also lessons on how that anatomy has changed through its evolutionary history. It could even, if desired in some curricula, include information on the relationships between phylogeny and the Earth's

[4] The Froggipedia app, mentioned above, includes animations of a maturing frog, showing its emerging legs and other changes.

[5] There are of course overlapping fields of study that investigate the evolutionary history of frogs and related species, examining fossils, performing carbon dating, and exploring DNA data to create fantastically—and at times beautifully—intricate family trees (e.g., Pyron and Wiens 2011; Feng et al. 2017; Portik et al. 2023).

geological history, and how these things are studied by scientists, such as through analysis of everything from DNA to the fossil record.

4.2.4 Variability

One important criticism of simulated dissection is that it does not possess the sample variability that comes with the dissection of dead frog bodies. Each dead frog is a different sample, and there may be pedagogical value in allowing students to see the individual differences, insofar as teachers facilitate this comparison in the classroom. Students can observe differences between male and female frogs, frogs of difference sizes, or ones with different stomach contents. In the case of simulated frogs, students all receive the same virtual instance, i.e., whatever version of the frog that programmers or manufacturers were able to create.

However, we can note that this lack of variability is not an inherent limitation of simulations in general, and computer simulations in particular. The flexibility of computer simulations actually opens up the possibility of a potentially large range of frog availability onscreen. We can imagine students toggling through multiple virtual frog samples to observe some of the differences possible between individual specimen. This changeability could extend to differences in frog species or perhaps different animal types entirely.

Another criticism of computer simulations is that they cannot reproduce an experience of felt consequences. According to this argument, since a computer simulation can simply be rebooted, and since in some cases the program may allow for actions (such as virtual cuts) to be reversed, students will not have the same experience of consequences for their choices and actions as they would in the dissection of a dead frog. Put another way, one virtue of the dissection of a dead frog, some defenders have argued, is that mistakes matter; a wrong cut cannot be undone. And, so the argument goes, there is a pedagogical value to the care and thought required for slicing well, and to the confrontation with irreversible repercussions for your choices. Computer simulation, it is argued, cannot really replicate this. (We can note as well that this is yet another argument made in favor of frog dissection that is not specific to this activity, but is one that is really instead arguing for the importance of hands-on educational experiences in general).

And it's a fair hit, I think, if perhaps a minor one. A simulation of course could in principle include a setting in which mistakes made are functionally irreversible. But such an option might be experienced as a bit arbitrary. This may simply be a drawback of computer-simulated dissection when compared with physical hands-on options like the dead frog or the physical simulation. However, it is only from within the assumptions of the dissection paradigm that the discussion on this point would end there. With such assumptions left behind, we can recognize the corresponding educational advantage of this issue for simulation: repeatability. The fact that a virtual cut could be performed over and over on a simulated frog has potential educational value. And while mistakes may be potentially resettable, there is potential educational value in the opportunity to make mistakes, reset, and try again.

The same way that computer simulations make it possible to easily correct for mistakes, they also make it possible to learn from these mistakes through repeated trial and error. Choices made during a computer-simulated dissection are potentially reversible. This means that multiple, quite different routes of dissection could be explored in a manner impossible in the case of the dead frog. We could imagine, for example, the possibility of a simulation that enables a view into the circulatory system, complete with simulated blood pumping through the veins onscreen. And we could imagine a temporary pause in that pumping as the student dissects the virtual frog's heart onscreen. Then, repairing the virtual heart, the blood onscreen could continue pumping again. By setting aside the expectation that a simulation must replicate the dissection of a corpse, this opens up opportunities to explore the dynamics of a functioning interior bodily system.

4.2.5 Media Integration

One key feature of computer-simulated dissection is its potential for integrating various media. When thinking about computer simulations in particular, it is possible to add different kinds of images and sounds to the experience. These can include everything from photography, to video, to computer-generated images and animation, to sound recordings. It is not uncommon for computer simulations of frog dissection to incorporate nature photography, animations of the frog cuts, 3-D images, and/or video of frogs in nature or of classroom frog dissections. Many of the above suggestions for expanding on these various forms of media integration are already a part of many contemporary dissection simulations.

Asked to reflect on the practice of slicing open frogs in high school biology classrooms, the philosopher of technology Don Ihde notes that this, "model for dissection is quite old, indeed, I would claim this dissection model is at least an eighteenth or nineteenth century model of biological knowledge based upon necessarily 'dead' objects" (2011, 203). There is something Victorian about the idea that a paradigmatic example of doing science involves snatching an animal from the wild, killing it, and fumbling around in its guts to see what's in there. Some aspects of contemporary scientific research of course continue in this manner. However, today there are a multitude of other modes of entirely normal scientific laboratory work that could also be modeled in this classroom activity. The flexibility of computer simulation enables possibilities for integrating parts of these other modes of scientific investigation.

Ihde continues,

> today's biological knowledge demands much, much more and focuses much more primarily upon a *dynamics*. How does the circulation system work, the nervous system work, digestion, or even the living dynamics of frogs catching prey or behave in other life situations? In these cases the scientific knowledge produced is often produced through contemporary *imaging technologies*. (2011, 204)

Many forms of contemporary scientific research include the interpretation of imaged readouts.[6] Scientific research across many fields utilizes data that take the form of images, and these are often displayed on screens. Medical imaging technologies, such as X-rays, fMRIs, and PET scans are used to investigate the interiors of bodies. These imaging perspectives on interior bodily anatomy are non-invasive. That is, they don't require cutting anything open to provide a view inside. They don't inherently require killing the specimen. Some imaging technologies provide a view of interior processes in motion, such as a sonagram displaying the movements of a developing fetus, or an fMRI displaying oxygen usage in the brain. The imaged data readout is of course not identical to the object imaged, the thing that has been transformed through the imaging process. That is, in the process of rendering an

[6] The generation, interpretation, and usage of images in practices of science and medicine is a major topic of study among scholars in fields such as philosophy of technology, philosophy of science, and Science and Technology Studies (e.g., Treichler et al. 1998; Coopmans et al. 2014; Carusi et al. 2015; Fried and Rosenberger 2021).

object of study—such as anatomical details within a living body—available to the scientist's perception, that thing is changed into something else. A 3-dimensional object in the world may be changed into a 2-dimensional picture onscreen. It may be isolated from the body, like an fMRI brain scan, with activity mapped onto a standard brain-shaped readout onscreen. Brain activity, or at least oxygen usage, is represented by false colors. And so on. Part of the job of many contemporary scientists is exactly the work of learning to interpret, decode, and find meaning within high-tech imaged readouts. That is, a central aspect of a lot of contemporary scientific research is the work of image interpretation. All of this could be useful to introduce to students. Computer simulated dissection offers the potential for integrating lessons on a variety of imaging data, enabling different ways to explore animal anatomy, and in many cases information and perspectives that are unavailable through dissection.

It is important to note that these kinds of imaged data often share the same interface format as computer simulations: screens. The onscreen format of computer simulations is, as noted throughout, conventionally cast as a chief limitation when held in contrast to the hands-on engagement with the dead frog. However, on this issue of potentially integrating imaged data, computer simulations are at distinct advantage. As a learning activity also already taking place onscreen, there is the possibility of straightforwardly integrating scientific imaging perspectives into the learning experience. Put another way, this represents an important opportunity to teach anatomy lessons in a way that includes distinctive perspectives on biological processes, and that connects to common contemporary high-tech scientific investigative techniques.

Simulations of dissection involve inherent tradeoffs, and this includes computer simulations of dissection. On the one hand, computer simulations are limited to providing only digitally mediated experiences onscreen. Yet, on the other hand, simulations maintain multiple educational benefits. These include a versatility that enables many potentially advantageous directions for expansion. They include the potential for displaying objects of study from various perspectives, including objects in motion (including the motion of internal bodily systems), as well as zoomed-in perspectives that allow close-up viewpoints, even, potentially, microscopic viewpoints. This versatility also includes the potential for integrating other ideas, lessons, and imagery into the simulation activity itself, including information on everything from the environment, to evolutionary history, to cell biology, to scientific imaging technologies.

The brainstorm here in this chapter should not be understood to be in any way comprehensive; just the opposite. I offer these ideas in an exploratory spirit. That is, these suggestions are merely intended to demonstrate that simulation expansion is possible, and that it could be potentially beneficial. The potential directions for expansion remain a wide open and uncharted territory.

References

Allchin, Douglas. 2005. "'Hands-Off Dissection?' What Do We Seek In Alternatives to Examining Real Organisms?" *The American Biology Teacher* 67 (6): 369–374.

Carusi, Annamaria, Aurora Hoel, Timothy Webmoor, and Steve Woolgar, eds. 2015. *Visualization in the Age of Computerization*. New York: Routledge.

Coopmans, Catelijne, Janet Vertesi, Michael Lynch, and Steve Woolgar, eds. 2014. *Representation in Scientific Practice Revisited*. Cambridge: MIT Press.

Dewey, John. 1938. *Experience and Education*. New York: Collier Books.

Feng, Yan-Jie, David C. Blackburn, Dan Liang, David M. Hillis, David B. Wake, David C. Cannatella, and Peng Zhang. 2017. "Phylogenomics Reveals Rapid, Simultaneous Diversification of the Three Major Clades of Gondwanan Frogs at the Cretaceous-Paleogene Boundary." *PNAS* 114 (29): E5864–E5870.

Fried, Samantha J., and Robert Rosenberger, eds. 2021. *Postphenomenology and Imaging: How To Read Technology*. Lanham: Lexington Books.

Goodall, Jane. (2002). "Forward." In *The Use of Animals in Higher Education*, ed. Jonathan Balcombe. Washington: Humane Society Press, pp. vi–vii.

Ihde, Don. 2011. "Dissection and Simulation: A Postphenomenological Critique." *Techné* 15 (3): 203–205.

Oakley, Jan. (2013). *Animal Dissection in Schools: Life Lessons, Alternatives and Humane Education* [Policy paper.]. Ann Arbor, MI: Animals and Society Institute. Available online at: www.animalsandsociety.org/research/digital-archive/

Orlans, F. Barbara. 1988. "Debating Dissection." *The Science Teacher* 55 (8): 36–40.

Portik, Daniel M., Jeffrey W. Streicher, and John J. Wiens. 2023. "Frog Phylogeny: A Time-Calibrated, Species-Level Tree Based on Hundreds of Loci and 5,242 Species." *Molecular Phylogenetics and Evolution* 188: 107907.

Pyron, R. Alexander., and John J. Weins. 2011. "A Large-Scale Phylogeny of Amphibia Including Over 2800 Species, and a Revised Classification of Extant Frogs, Salamanders, and Caecilians." *Molecular Phylogenetics and Evolution* 61: 543–583.

Treichler, Paula A., Lisa Cartwright, and Constance Penley, eds. 1998. *The Visible Woman: Imaging Technologies, Gender, and Science*. New York: New York University Press.

CHAPTER 5

Conclusion: Save the Frogs

Abstract The previous chapters have developed what can be called the materiality argument against frog dissection. This argument has proceeded through the identification and criticism of assumptions often made about animal dissection and its alternatives that can make dissection appear to be the better educational option for precollege classrooms. With such biases set aside, the materiality argument continues on to consider the often under-recognized educational limitations and problems of animal dissection, as well as the often under-explored potential for expanding dissection simulation beyond a primary aim of mimicking the cutting up of a dead animal. This final chapter reviews this education-focused argument and combines it with the larger ethical and environmental case against classroom animal dissection.

Keywords Frog dissection · Educational technologies · Biology education · Simulation · Animal welfare · Humane education · Technological mediation · Philosophy of technology · Postphenomenology

In some ways, this is an odd historical moment to make a case against frog dissection. Even more so, it may be an odd time to extoll the benefits of classroom computing. Looking back, this past generation has seen an invasion of computing into many classrooms, building on the influx seen

R. Rosenberger, *Against Frog Dissection*,
https://doi.org/10.1007/978-3-032-14346-4_5

in the generations before (with of course class-marked differences across regions and school districts). Any number of advocates and purveyors for classroom computing have pointed to all variety of touchscreens, tablets, apps, websites, and adaptive testing programs as the latest cure-all for educational challenges. Parents and students fumble with various platforms and passwords. Critics bemoan excessive screentime, smartphone and social media dependence, and allegedly shattered attention spans. It would be easy to dismissively slot this book's endorsement of simulated frog dissection into a general trend—for better and for worse—toward a reliance on classroom computing.

Even more, it is possible that the coming generation of educational advances and concerns will at least in part be led by the growth of generative artificial intelligence. AI raises challenges for teachers' and students' conceptions of education itself. It could continue to change what students think about learning and what teachers think about teaching, especially in the case of widely available accesses to large language model text generation. In the worst case scenario for educators, some students may wonder: Why learn to write when AI will do the writing for you? Why learn to read well when AI will provide you an instant summary? Why sharpen your critical thinking skills when AI can generate a list of reasons in an instant? On top of these kinds of fundamental questions about the nature and purpose of education introduced by AI, there are also immediate practical pedagogical concerns: AI makes it easy to cheat on many conventional assignments.

In this context, we can expect there to be some periods of pushback against this inundation of screens and computing and AI into classrooms. My own personal sentiments tend to be with these kinds of critics. Against a tide of increased reliance or computing and AI, we can expect there to be occasional resurgences of calls for students to have real experiences in the classroom, hands-on experiences, face-to-face experiences, interactions, and explorations with the stuff of the world. Again, my own personal sentiments tend to be in support of these kinds of calls.

It is precisely on this point that a danger can be identified regarding our arguments here over frog dissection. We should not allow a sentiment of support for hands-on, in-person learning experiences *in general* to smuggle in support for *the particular* hands-on learning activity of frog dissection. *We cannot allow the debate over frog dissection and its alternatives to become a proxy for a debate over hands-on educational activities generally.* We should not allow the debate over frog dissection to be subsumed under larger arguments for and against classroom computing in general. As I have tried to show throughout this book, the frog dissection activity and its alternatives should be evaluated in their own terms, with consideration of their own particular educational benefits and drawbacks.

Part of the argument of this book has thus been an attempt at intuition management. The move to explicitly conceive of these activities as concrete material practices is part of this attempt. I have argued that remembering that both the frog dissection activity and its alternatives are themselves specific material things can help us to keep track of our intuitions here.

One way to draw out these intuitions is to think explicitly about how it is easily possible for a person to simultaneously agree that:

1. Hands-on, exploratory educational experiences with physical objects are generally important for education, and should be implemented in the classroom whenever possible.

   and

2. Alternatives are preferable in the specific case of frog dissection.

Any sympathy we have for the creation of experiences in the classroom in a progressive, Deweyan sense should not somehow commit us to support the particular experience of cutting up a dead frog. The frog dissection debate is not an avatar for debates over progressive education.

Another way to draw out these intuitions is to point out their role within what I have been referring to here as the dissection paradigm. The identification and interrogation of the dissection paradigm—these alleged patterns of assumptions often made regarding frog dissection and its alternatives—is another attempt at intuition management. We can observe many things to be associated with the frog dissection activity in those places in the world where it is common: that it's a rite of passage, that it's a tradition, that it calls for a mature or perhaps even macho attitude, that it's scientific and objective, that it represents an encounter with nature, that it's the real thing. There are reasons for thinking these things about frog dissection, but they are not uncomplicated. They are fraught and contestable. By identifying these things as features of a pervasive and unreflexive way of thinking, we can perhaps cut off some of their rhetorical power. Yes, there are genuine points to be made regarding the frog dissection activity's status as a real thing, or as a natural thing, or as a hands-on experience, etc. But we should not allow the kind of lazy thinking often seen within the dissection paradigm in which these features are assumed to be decisive. They are not.

While there is a genuine debate to be had about educational animal dissection, my claim has been that much of the actual discussion in practice occurs within a strongly established context of biased background assumptions, preconceptions, and intuitions. As a project of intuition management, much of the work of this book has been to find concepts, points, and examples that draw out and undercut those biases.

5.1 Technological Optimism, Pessimism, and Neutrality

These continual concerns about AI point to a broader issue: techno-optimism. As noted toward the start of chapter three, there are many tendencies in the ways we conceive of technology that could potentially weaken our critical thinking. For example, it is best to avoid "deterministic" assumptions about technology, whether we're being overly optimistic or overly pessimistic about a device's expected impact on the world. That is, we should check ourselves for unreflexive "optimistic" or "utopian" assumptions that new technologies will always necessarily make things better, or that their rollout and consequences will go exactly as we plan. And we should check ourselves as well for unreflexive "pessimistic" or "dystopian" assumptions that technologies can only make things worse, or less authentic, or less real. In addition to all this, we need to keep in check another opposite but also misleading tendency in our thinking on technology: that it is inherently innocent. That is, we should check ourselves for unreflexive "instrumentalist" assumptions that technologies cannot influence users in any way, or that they cannot ever play any role in shaping our thoughts, actions, and choices. Arguments about educational frog dissection and its alternatives, including those developed through this book, swim within the crosscurrents of our general intuitions about technology.

For example, some of us may harbor a general enthusiasm for educational technologies. Others of us may maintain a general skepticism about them. I've been attempting to argue in this book that we keep these general intuitions in check in the specific case of frog dissection and its alternatives. Let's evaluate frog dissection and its alternatives in terms of the specifics of their educational advantages, disadvantages, and problems. And let's not let these evaluations get swept up in our general impressions—positive or negative—for classroom computing.

If the preceding chapters have developed an argument *against* frog dissection in childhood education, then they also make up an argument *in favor* of specific simulated alternatives, and in particular educational computing alternatives. The final part of the fourth chapter brainstormed various ways that dissection simulations could be expanded beyond the task of merely imitating cutting up a dead frog, with suggestions including everything from microscopic vantage points, to information about frogs' lifespans and environment, to imaging readouts of frog bodies, among many other potential educational directions for development. If we want to avoid an instrumentalist intuition—one which mistakes technology as somehow always fully innocent—then we shouldn't assume that either the frog dissection activity itself, or simulated alternatives, are neutral actors within our human-technology relationships with the world. Since this is so often ignored in discussions on frog dissection, much of this book has been about exposing, identifying, and analyzing the non-neutrality of the technology of the dead frog.

However, we must also not forget the non-neutrality of the expansions to dissection simulation that have been suggested above. This does not reduce to the non-neutral move from the hands-on faux-experimental context of the dead frog to the virtual context of the computer screen. In addition, any expansions made to dissection simulation would also of course come with their own tradeoffs. For example, the work of expanding simulation and getting it into classrooms could be financially costly, although of course this can be the case as well for frog dissection and physical dissection simulations. There are also opportunity costs to consider. As we've brainstormed, there are many potential educationally beneficial directions that dissection simulation could be expanded beyond merely looking at anatomical parts, e.g., into information about frog habitat, or about microscopic biological processes, or about evolutionary history, etc. However, the expansion of a particular direction or directions may need to come at the expense of others. Class time itself is also of course not an unlimited resource. Expanding the kinds of lessons addressed by dissection simulation could cut into time for other things, or reorganize how things are taught. (This is of course also the case for maintaining the status quo. The continued use of animal dissection in class comes with opportunity costs. Or as Orlans puts it, "The current over-reliance of some teachers on dissection comes at the expense of other laboratory studies" (1993, 200)). In addition, as with any new or existing classroom technologies and activities, there are issues of teacher training. Computer programs must be accessed, or licensed, or downloaded, and however else installed, or opened, or otherwise made available students.

I bring up these points regarding some of the basic realities of implementing expansions of educational dissection simulations in part because I want to keep the discussion on these things at a concrete level. If we can set aside the biased preconceptions of the dissection paradigm, then we can take an honest look at many things, including the well-known advantages of frog dissection, as well as its educational limitations and problems, and also the advantages and disadvantages of simulated alternatives, including their strong pedagogical effectiveness, their avoidance of many of the problems and limitations of dissection, and their yet unmet potential for educational expansion. Within this larger context of evaluation, simulated dissection even more clearly stands out as the preferable option in terms of educational benefit.

5.2 Concrete Materiality and the Ethics of Frog Dissection

The argument developed throughout this book—the materiality argument, as I have called it—draws out all of these dynamics. And it can be used to support the moral and environmental arguments against frog dissection, and to support those in favor of its alternatives.

In this recasting under the materiality argument, both frog dissection and its alternatives come out looking a lot different. Let's review.

When we approach the frog dissection activity as concrete material practice, it can be recognized for its educational benefits, but also its limitations and problems. We can acknowledge the distinctive educational advantages of slicing open and exploring the body of a dead frog, including the hands-on physical engagement, and the sense of discovery that accompanies the exploration of its insides. With its physical resistance, smells, and queasiness, the frog dissection activity can also be acknowledged to be especially impactful and potentially memorable. However, we must correspondingly recognize the many relevant lessons and educational connections that the specificity of the dead frog disallows. These limitations include, for example, the exclusion of relevant lessons on animal habitat, as the frog has been removed from its environment as part of the classroom faux-laboratory posturing. They also include the prohibition of any exploration of frog anatomy as dynamic bodily systems, since the already dead frog lies there immobile. In addition to these and other limitations identified in the previous pages, we must also recognize important ways in which the specificity of the frog dissection activity makes it educationally problematic: it implicitly imparts a morally questionable lesson on the value of animal life, as well as a problematic lesson on desensitization and science. As the frog has been killed for this activity, and as it will be tossed in the garbage or otherwise left behind at the end, frog dissection teaches a low regard for animal life, implying it to be something not worth as much as an afternoon lesson that could have been learned with equal effectiveness through other means. And, in its posturing as a scientific experiment, the frog dissection activity teaches that the work of science involves compartmentalizing any disgust, moral or otherwise, that you may feel in practice.

In addition, approaching frog dissection as a concrete material practice highlights the various associations it has gathered within these places on the planet where it remains an expected part of a child's school years. Following others, I've been referring to this broad set of associations here as the dissection paradigm. Under the dissection paradigm there are a number of questionable things that are assumed about the frog dissection activity and its alternatives, assumptions that I suggest have the effect of making the frog dissection activity appear to be the obvious option to prefer. For example, one feature of the thinking under the dissection paradigm that we've been wrestling with throughout this book is that the frog dissection activity itself simultaneously serves two roles: (1) as one educational option, and (2) as the very standard or the ideal against which other options are measured. Insofar as we allow the technology of the dead frog to serve this role as the standard of comparison, we are cut off from critically analyzing the frog dissection activity and assessing its limitations and faults. Instead, we are left only to evaluate how well or poorly the alternatives measure up.

We can also think about how these technologies help to shape our conception of what's "real" and what's not, at least within these moments and spaces of the classroom. Discussions on frog dissection and its alternatives are often cast in terms of real and virtual options: the dead frog is a physical object there in the classroom, hatched and grown through natural processes, and lying there waiting to be manually manipulated and mutilated; simulated alternatives instead present a artificial thing, whether it's a human-made physical simulation, or a human-programmed computer simulation onscreen. These are relevant distinctions to recognize. However, under the dissection paradigm, these distinctions get fetishized. That is, within discussions on educational dissection, these distinctions can misleadingly be cast as if they are mutually exclusive, as well as more important than they actually are. An understanding of the dead frog as the real thing, and merely the real thing, can have an effect of cutting off critical analysis of the frog dissection activity in terms of the actual ways that it has passed through human hands, ways that are both educationally and ethically relevant.

We can also approach alternatives to frog dissection as concrete material practices. One of the central arguments coming out of the previous pages is that if we can manage to pull dissection simulations out from under the shadow of the frog dissection activity, then we can better analyze these simulations in terms of their own particular educational advantages and setbacks. Rather than only consider simulation in the educational terms set by the frog dissection activity, we can think about the concrete material practices of dissection simulations in their own terms, i.e., with regard to the specific educational experiences they create in the classroom, and with regard to their own educational potential. In the previous pages, I attempted to brainstorm some directions that simulations could be expanded. Ultimately the point here is to demonstrate *that* dissection simulation can and should be expanded. Educators and developers could take this in many directions, and surely can come up with more and better avenues for educational expansion than I have here. Considering the strength of the grip of the dissection paradigm in those places where it predominates, this means that a project of expanding dissection simulation beyond the limits set by the dissection paradigm takes more than only creativity, more than pedagogical insight, and more than the work of simulation design and development. It takes the work of overturning the paradigm.

I offer the materiality argument as a kind of support to bolster the arguments against frog dissection. Critics argue that it is immoral for frogs to be killed merely for the sake of this activity, especially in the face of the educational effectiveness of alternatives. Since the frogs for dissection are often captured in the wild, critics also contend that the practice of frog dissection enacts an unacceptable environmental cost. And it is argued that the frog dissection activity implicitly imparts an immoral lesson to students that animal life has little value. These moral and environmental points are strengthened by the addition of the materiality argument. This strengthening occurs in part through the recasting of the educational counterarguments. When—following the materiality argument—we approach the frog dissection activity and its alternatives not as ideals or proxies but as actual concrete material practices, we can better reflect on frog dissection's under-acknowledged limitations and problems, as well as dissection simulation's yet untapped potential.

With a conception of these concrete material practices on hand, we can see even more clearly that many of the common refrains repeated to dismiss the educational potential for dissection simulation are not enough to defend the frog dissection activity against these moral concerns.

It is not enough to merely state that a real frog is better than a virtual one onscreen, as we see this claim to be more complicated and fraught that it appears on the surface. It is not enough to state that frog dissection is a hands-on activity, as there are many other hands-on learning experiences we could be arguing for instead, rather than this problematic one. And it is not enough to merely claim that simulation cannot ever reproduce the same exact experience of cutting up a dead frog, as that is not the only or best standard of comparison.

Let's allow dissection simulations to expand. Let's stop dissecting dead animals in childhood and young adult education. Let's save the frogs.

Reference

Orlans, F. Barbara. 1993. *In the Name of Science: Issues in Responsible Animal Experimentation*. New York: Oxford University Press.

Index

R. Rosenberger, *Against Frog Dissection*,
https://doi.org/10.1007/978-3-032-14346-4

FSC
www.fsc.org
MIX
Papier | Fördert
gute Waldnutzung
FSC® C083411

Kolibri 360 GmbH
Ferdinand-Jühlke-Straße 7
99095 Erfurt, Deutschland
produktsicherheit@kolibri360.de